D0384663

UNCORKED

UNCORKED

The Science of Champagne

GÉRARD LIGER-BELAIR

PRINCETON UNIVERSITY PRESS *Princeton & Oxford*

Published by Princeton University Press, 41 William Street, Princeton,
New Jersey 08540
In the United Kingdom: Princeton University Press, 3 Market Place,
Woodstock, Oxfordshire 0X20 ISY

LIBRARY OF CONGRESS CATALOGING-IN-PUBLICATION DATA
Liger-Belair, Gérard, 1970–
Uncorked : the science of champagne / Gérard Liger-Belair.
p. cm.
Includes bibliographical references and index.
ISBN 0-691-11919-8 (cloth : acid-free paper)
1. Champagne (Wine). I. Title.
TP555.L54 2004
641.2′224—dc22 2004044604

British Library Cataloging-in-Publication Data is available
This book has been composed in Perpetua and Centaur
Printed on acid-free paper. ∞
pup.princeton.edu
Printed in the United States of America
1 3 5 7 9 10 8 6 4 2

To my father, Jacques

CONTENTS

o Introduction 1

oo The History of Champagne 7

ooo The Making of Champagne 19

oooo A Flute or a Goblet? 31

ooooo The Birth of a Bubble 37

oooooo The Bubble Rises 59

ooooooo The Bubble Bursts 85

oooooooo Afterword: The Future of Champagne Wines 133

Glossary 143

Bibliography 145

Acknowledgments 148

Index 149

INTRODUCTION

Champagne has launched thousands of ships, toasted countless weddings, and inaugurated billions of New Year's parties throughout the world. Almost everyone—certainly everyone reading this book—has an interesting story to tell that includes a bottle of champagne. So it seems best to start this book with a story of my own.

I'm a physicist. What's a physicist doing writing about champagne? Well, the story begins on a summer afternoon when I was a student in the midst of finals and thought it would be a good plan before getting on with studying to stop somewhere on the way home to have a beer. Now keep in mind that this is a *physics* student stopping somewhere to have a beer, and besides, I had a predilection for fluid dynamics and, on the side, photography. The sunlight, the hot day, the alcohol, my studies, and the thought of actually trying to study more later on all helped to focus my attention on the golden bubbles rising up through the beer and along the sides of the glass in front of me. . . . I sat, mesmerized. I thought, Effervescence really belongs to that category of daily phenomena that naturally engage the imagination; I could just as well be watching the clouds in the sky, flames popping in a fireplace, or waves breaking on a beach. I suppose that I could have just left it at that, but I wanted to know more. I wanted to see closer, get my camera. I suddenly realized, in a flash of free association (and slight intoxication), that I wanted to study carbonated beverages.

A year later I had a master's degree in fundamental physics and was still a bubbles addict. I bought a secondhand macrophotographic lens and, over the holidays, started taking photographs of bubbles rising in a glass filled with soda. I spent whole nights developing film and enlarging pictures in my bathroom, which—to my girlfriend's chagrin—I had transformed into a makeshift darkroom. Six months and many sleepless nights later, I mailed some of my best shots to the research department of Champagne Moët & Chandon, along with some initial scientific observations describing what was occurring in the photographs.

My grand plan was this: Champagne makers such as Moët & Chandon sold 262.6 million bottles in 2001—the equivalent of about $3 billion in sales. For an industry that banks so much on bubbles, finding ways to better understand the bubbling process and eventually to improve the beverage's hallmark fizz seemed like a smart idea. Department head Bruno Duteurtre asked me to come to Moët & Chandon's headquarters in Epernay (the capital of Champagne wines) to lay out my photographs, thoughts, and research plans. The people I met with were captivated by the idea of this research, and a few weeks later I moved from Paris to Reims to begin my dissertation in the Laboratory of Enology at the University of Reims. With colleagues from both the university and Moët &

Chandon, I started my official investigation into the physical chemistry of champagne bubbles.

As Dr. Harold Edgerton, a revolutionary in the development of high-speed and stop-motion photography, once said, “The experience of seeing the unseen has provided me with insights and questions my entire life.” This sentiment exactly captures the heart of the matter, the reason for this book, and the answer to my earlier question. Champagne is a wonderful drink, one that mysteriously manages to capture an incredible amount of festivity, elegance, and sensuality in every glass. A lover of champagne—either a guest at your next New Year’s party or a connoisseur at a banquet—certainly can drink a glass and enjoy it with great pleasure. These two lovers of champagne may have different vocabularies to describe what they find pleasurable about the wine, but it’s very likely that neither knows nor can even begin to imagine all that is happening inside the flute in his or her hand. Mainly, this is so because some of the most interesting and beautiful events in a glass of champagne are invisible to the unaided eye. With a special lens on a camera, however, we can capture photographs of the bubbles in champagne. We can study those photographs, and with the help of a little physics know-how, understand how the bubbles act as the vehicles for taste, scent, that lovely popping sensation on your tongue and the tip of your nose—in general, the pleasure we all know, love, and have come to expect from a glass of champagne.

Like music, like poetry, and like the personalities of people, our physical environment—animate or inanimate, natural or manufactured, wild or civilized—has beauty to be appreciated both on a superficial level and at a deeper level of structural understanding. It behooves an admirer to examine an object of beauty closely so as to enrich his or her experience and appreciation of its charm. As a physicist—not a connoisseur, not a poet—I obviously have a different vocabulary and level of expertise to make use of than others might employ in the enthusiasts' literature when they describe the experience of drinking a glass of champagne. However,

I hope to go beyond vocabulary and offer a unique experience to you, regardless of your background or expertise: to see what the unaided eye cannot see and to observe closely some of the building blocks of beauty—the physics and the chemistry of that which gives champagne its charm and gives us our pleasure. I hope that your enjoyment of champagne is only enhanced by this guide that a scientist has created to the lovely physics of the bubbles that sparkle within it.

THE HISTORY OF CHAMPAGNE

A Closer Look: The Sparkling of the Wine

Pop open a bottle of champagne, pour yourself a flute, and observe what happens in the small space inside the glass. Bubbles form on several spots on the glass wall, detach, and then rise toward the surface in elegant trails, like so many tiny hot-air balloons. Listen carefully also. When they burst at the surface, the bubbles make a crackling sound and produce a cloud of tiny droplets that pleasantly tickle the taster's nostrils. This process—effervescence—enlivens champagne, beers, and many other carbonated beverages. Without bubbles, beer would be flat, and champagne and sparkling wines would be unrecognizable.

Critics judge a champagne by, among other qualities, its bubbling behavior. Small bubbles rising through the liquid—as well as the collar of bubbles, or *collerette*, at the periphery of a flute of champagne—are the hallmarks of this festive wine. Even if there is presently no scientific evidence to correlate the quality of a champagne with the fineness of its bubbles, people nevertheless do make a connection between them, and it is often said, "The smaller the bubbles, the better the wine." One possible origin of this saying is likely aesthetic; small bubbles rise at a more leisurely pace than larger ones do and consequently create the wine's characteristically lingering sort of effervescence and delicate inner glow. However, another reason for this saying (described in more detail in Chapter 6) is that older (and often better-quality) champagnes lose some of their carbon dioxide during the aging process, and consequently, they present smaller bubbles when their bottles are opened for tasting than do younger (and often lesser-quality) champagnes. Over time, as wine drinkers took note of this phenomenon, they eventually may have made the generalization associating the size of the bubbles with the quality of the wine and coined the popular phrase.

A Bit of History

So how did all this bubbling start? The history of this unique sparkling wine, still rather controversial, begins at the Abbey of Hautvillers, one of the oldest Benedictine abbeys in the world. The abbey is situated high up in the Valley of the Marne, a northeastern region of France around 150 kilometers from Paris and not far from the cathedral town of Reims, where the kings of France were crowned. The Champagne region was then and is now one of the select high-latitude locations where it is possible to grow good grapes for winemaking.

Wines from the Champagne region were long-time favorites of the kings in Paris. Until around 1500, Champagne's wines were still *sans* bubbles and competed largely with wines from Burgundy, often successfully because it was much easier to get them down the Marne River to Paris. In the late 1400s, though, the weather got colder in Europe. This change in temperature was to make huge waves in the winemaking industry.

Temperatures plunged suddenly in the northern hemisphere at the end of the fifteenth century. Large bodies of water across Europe froze, including such landmarks as the Thames River and the canals of Venice. At vintage time in the Champagne region the weather suddenly was much colder than normal. In winemaking, the yeast that was placed on the grape skins to convert the sugar in the pressed grape juice into alcohol in the warm months did not have sufficient time to do its job because the sudden cold abruptly stopped the fermentation process. When spring came, fermentation began again, but this time within the casks and containers in which the juice had been bottled. This second fermentation produced carbon dioxide in excess, which became trapped inside the containers and created a slight fizziness. . . . Champagne was born.

The French aristocracy hated this "fizz" and considered it a sign of poor winemaking. The market for Champagne's wines collapsed progressively and finally was lost entirely to Burgundy. Champagne wines were to suffer two hundred years of hard times before the Catholic Church—which had a major interest in Champagne's vineyards and was experiencing revenue losses—decided to tackle the problem. In 1668, the church assigned a twenty-nine-year-old monk by the name of Dom Pierre Pérignon the task of getting rid of the bubbles and producing still (nonsparkling) wines such as those the winemakers of Champagne had made previously with great success. Dom Pérignon was appointed the new cellar master of the Abbey of Hautvillers, and he proceeded to develop several empirical methods to reduce (but not completely prevent) bubbling.

In the meantime, while Dom Pérignon struggled to debubble the abbey's wines, tastes began to shift. "Sparkling wine" was picked up as a fad and suddenly became fashionable in high society. In England, a newly sophisticated and somewhat decadent society formed under the reign of Charles II (who ruled from 1660 to 1685 in the time that's also remembered as "Merry Olde England"), and it included some fans of sparkling wine. In fact, it seems that sparkling wines had existed already in England in some form at least two or three decades before being produced for consumption in the Champagne region of France. In December 1662 (six years before Dom Pérignon became the cellar master of the Abbey of Hautvillers), an Englishman by the name of Christopher Merret presented a paper on making sparkling wine to the newly formed Royal Society of London.* Independent of the French "discovery"

* The relevant page of Merret's original paper, "Some Observations Concerning the Ordering of Wines," was first published by Tom Stevenson in *Christie's World Encyclopedia of Champagne and Sparkling Wine* (Bath, England: Absolute Press, 1998).

of champagne induced by the sudden cold snap, Merret had found that simply adding sugar to Champagne's wines made them effervescent and increased their alcohol content, and he stated as much in his paper. Consequently, in England, many noblemen would order still (nonsparkling) wine in casks, add a dose of sugar, and then bottle it. The English were inveterate improvers, but in this case, as Merret points out, they added the sugar not for taste but for the express purpose of making the wines sparkling and to increase their alcohol content.

Back in France, members of the Royal Court at Versailles under Louis XIV also began to appreciate bubbles in their wine. At the end of the seventeenth century, therefore, Dom Pérignon was ordered to reverse his efforts and to devise methods to *increase* the bubbling in the wine. This must have made him very happy because purportedly—when he first tasted a wine that had become effervescent accidentally—he yelled to his fellow monks, "Come quickly, brothers, I am drinking stars!" Though Dom Pérignon was not the one to introduce bubbles to a still wine, he went on to develop new ways to increase the bubbles in the wine and new techniques that led to the development of champagne as we know it today.

In particular, he developed an efficient cork (inspired by visiting Spanish monks with cork stoppers in their water bottles) for use as a bottle stopper instead of the typical rags or wooden pegs. Use of the hermetic cork was instrumental in maintaining a large quantity of bubbles in the sparkling wine. However, this same cork produced some disastrous effects. While undergoing the second fermentation, champagne requires a container that can withstand several atmospheres of pressure when sealed. Otherwise, the internal pressure generated by the carbon dioxide builds up inside the sealed container and eventually causes it to explode. Since French glassmaking technology was still not highly developed at this time and thus bottles were unable to withstand such high internal pressures, many bottles of champagne did blow up from either improper corking, internal pressure, irregularities in the glass, or a combination of all these factors. It was so dangerous to deal with champagne bottles, in fact, that buyers had to wear facemasks when walking through a champagne cellar. Often more than half the bottles exploded before ever reaching their ships, parties, or toasts.

The English, it turned out, also were having a similar problem making strong enough glass to withstand the high internal pressures of their sparkling wines. However, in the early seventeenth century, glassmaking technology made huge progress in England due in large part to an admiral named Sir Robert Mansell. Sir Robert had observed that forests and natural wood supplies were being devoured rather quickly for fuel—and not in small quantities by glassblowers—and grew seriously concerned about the future of British shipbuilding. He expressed his anxieties to King James I and succeeded in persuading the king to forbid glassblowers to fuel their kilns with wood. As a result of this decree, new coal-burning furnaces were developed—and these kilns were able to reach such high temperatures that the internal structure of glass was modified in such a way that it could be used to produced darker but much stronger bottles, which were ideal for champagne storage.

Dom Pérignon—said to have been endowed with an amazing sense of smell and a discerning palate—also went on to invent and develop the art of *blending*, which is really the essence of winemaking in Champagne nowadays. Blending consists of mixing different base wines from various grapes, vineyards, and years to create another wine that is superior to any of its parts. By some accounts, even when he was blind toward the end of his life, Dom Pérignon was able to identify the grapes *and* the vineyard from which a wine was made. Dom Pérignon also developed new techniques to press black grapes to yield a white juice and improved clarification techniques to create the most brilliant wines ever produced at that time.

In 1715, after forty-seven years as cellar master, the "father of champagne" died. A little over a decade later, King Louis XV, great-grandson and successor of King Louis XIV, recognized the importance winemaking could have as an industry for France and issued a decree in 1728 granting the City of Reims exclusive permission to trade in bottled champagne. In 1729, Nicolas Ruinart founded the first recorded champagne house, and in 1743, Claude Moet founded what was to become the largest champagne house in existence today, the House of Moët & Chandon. Moreover, since the basic principles Dom Pérignon established to make champagne and sparkling wines are still used throughout the world today, it should come as no surprise that a large statue of Dom Pérignon adorns Moët & Chandon's central office. However, in 1936 the company paid even greater homage to this man of influence by launching Moët & Chandon's special and most prestigious blend, Dom Pérignon.

○ ○ ○

THE MAKING OF CHAMPAGNE

The modern production of champagne is not far removed from that developed empirically by Dom Pérignon. This method, known as the *French Method*, is also used outside the Champagne region; wines produced as such are labeled *Méthode Champenoise* or sometimes *Méthode Traditionelle*. Indeed, most American and Australian makers of sparkling wines use this method to create their champagnes. The *Méthode Champenoise* involves several distinct steps.

A First Alcoholic Fermentation

Three types of noble grapes* are grown in the 75,000 acres of the Champagne vineyards: *chardonnay* (a white grape), *pinot meunier* (a dark grape), and *pinot noir* (a dark grape). Usually around mid-September the grapes harvested from these vineyards are pressed to make a juice called the *grape must*. After pressing, the must is transferred into an open vat, where yeast (a kind of fungus known scientifically as *Saccharomyces cerevisiae*) is added.

Generally speaking, the key chemical reaction in winemaking is alcoholic fermentation—the conversion of sugars into ethanol (the intoxicating agent in liquors) and carbon dioxide (the source of effervescence) by yeast (Figure 1). Observation of the fermenta-

* The term *noble grapes* is used to refer to the grape varieties well known to produce the world's great wines: sauvignon blanc, chardonnay, riesling, merlot, pinot noir, pinot meunier, and cabernet sauvignon.

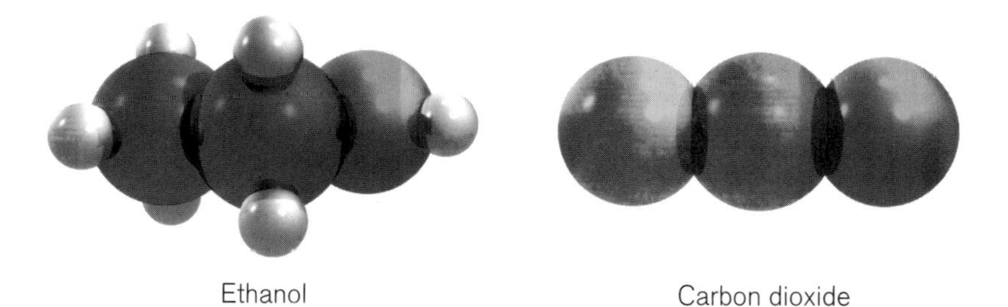

Figure 1 The fermentation of sugar by yeast produces ethanol and carbon dioxide molecules.

tion on the ground of ripe fruits that dropped from their trees probably first led to the discovery of alcohol by our forefathers, and as mentioned earlier, the English documented their knowledge of the fact that adding sugar to a cask of wine and recorking it would work to increase the wine's levels of alcohol and effervescence. However, French chemist Joseph-Louis Gay-Lussac first described the process of fermentation in scientific terms in 1810 when he demonstrated that glucose is the basic starting material for producing ethanol. The manner in which yeast contributes to the fermentation process was not clearly understood until 1857, when the famous French microbiologist Louis Pasteur defined the following equation and discovered that not only does the fermentation process not *require* any oxygen but also that alcohol yield actually is *reduced* by its presence:

$$C_6H_{12}O_6 \rightarrow 2CH_3CH_2OH + 2CO_2$$

In simple language, this equation means that glucose ($C_6H_{12}O_6$) is consumed by yeast and then broken down into ethanol (CH_3CH_2OH) and carbon dioxide (CO_2). The amount of alcohol generated by this first fermentation is about 11 percent. This percentage is the maximum amount of alcohol produced before all the yeast cells are killed by the alcohol concentration. At this step, "champagne" is still actually a noneffervescent white wine because the carbon dioxide produced during the first fermentation is allowed to escape into the atmosphere.

The Art of Blending

Because it is rare that a single wine of a single vintage from a single vineyard and grape variety will provide the perfect balance of flavor, sugar level, and acidity necessary for making a fine champagne, winemakers often will mix several different still wines. This is called the *assemblage* (or *blending*) step, and it is carried out directly after the first fermentation is complete. Blending is considered to be the key to the art of champagne making. A cellar master sometimes will blend up to forty different wines from various grape varieties, vineyards, and years to produce one champagne. In a way, the cellar master is the architect of the sparkling wine. This blending requires considerable insight because it is extremely difficult to predict the final result of blends that will be consumed years later. The blending of still wines originally made from the three kinds of grapes (with the right proportions supplied by an enologist*) forms a base wine that will then undergo a second fermentation—the key step in producing the "sparkle" in sparkling wine.

The *Prise de Mousse*: The Second Fermentation

Once the base wine is created, sugar (about 24 grams per liter), yeast, and yeast nutrients are added. The entire concoction is put into a thick-walled glass bottle and sealed with a cap. The bottles are then placed in a cool cellar (12 to 14° C), and the wine is allowed to ferment slowly for a second time, producing alcohol and carbon dioxide again. Those 24 grams of sugar per liter of wine will produce around 12 grams of carbon dioxide and a gain of alcohol of around 1.5 percent. This time, however, the bottles are sealed so that the carbon dioxide cannot escape and thus remains

* An *enologist* is a person who is in charge of the entire process of winemaking at a vineyard (from grape must to bottling). *Enology* is the science and study of wine and winemaking. In comparison, the wine waiter, or *sommelier*, is the person in charge of wines and wine tasting at a restaurant.

in the wine. The carbon dioxide is not yet in the form of bubbles but rather in the form of carbon dioxide molecules dissolved in the wine. During this second, slow fermentation process, the carbon dioxide dissolved in the wine and the gaseous carbon dioxide under the cork progressively establish equilibrium (an application of Henry's law*). Once this prise de mousse is over, the pressure under the cork is about 6 atmospheres (the same pressure you would feel if you were under about 50 meters of water).

Aging

As the second fermentation proceeds, yeast cells die; after several months, when all the yeast cells are dead, fermentation is complete. Under the cork, the gaseous pressure has reached up to 6 atmospheres, but the champagne continues to age in a cool cellar for at

* Henry's law states that the pressure of a given gas above a solution is proportional to the concentration of the gas dissolved in the solution

least nine months and sometimes for several years. This process of *aging* allows the development of the so-called champagne bouquet. During this period, yeast cells split open, and their insides spill into the solution in a process called *yeast autolysis*, imparting complex and yeasty flavors to the champagne. A "toasty" flavor, very much sought after, comes from further chemical breakdown of the dead yeast cells. The longer the champagne ages, the richer this flavor becomes, and the more a champagne is valued. The best and most expensive champagnes are aged for five or more years.

Riddling and Disgorging

After the aging is complete, the champagne maker still has one problem remaining: how to get the dead yeast cells out of the bottle without losing the precious carbonation. The procedure developed to do so is known as *remuage*, or *riddling*. The process of riddling is important both for aesthetic reasons and for considerations of taste. With regard to the look of the champagne, if the dead yeast cells were not removed, you would find your glass of champagne to be distinctly cloudy, and certainly, the attractive sparkle of the drink would not shine through so delightfully. This may be reason enough for riddling, but there is also the matter of taste. As mentioned earlier, some champagnes are allowed to age for a long time before they go through the riddling process, whereas some are not. This is a decision that is made with the aging potential of the wines in mind. A wine of excellent potential likely will develop a more complex flavor and tempting bouquet by undergoing a longer aging prior to riddling; in contrast, a wine of lesser potential may be riddled while it is still young, with the knowledge that it is unlikely to improve after a certain amount of aging in contact with the yeast cells.

The process of riddling itself involves placing the bottles in specially designed racks that keep their necks tilted downward. This tilting forces dead yeast cells into the neck of the bottle. Twice a day, the bottles are *riddled*, meaning that someone lifts each bottle out of the rack slightly, turns it a quarter of a turn, and then thumps it back down into the rack. In ancient times, this process was done by hand and required great skill. A good *riddler* typically handled 20,000 to 30,000 bottles a day. Nowadays, automatic machines handle the job, and the bottles are turned completely upside down.

The bottles then undergo a *dégorgement*, or a disgorging process. In *dégorgement*, the necks of the bottles are frozen, creating a small ice plug at the top of the bottle that traps the sediment of dead yeast cells. The bottle caps are removed, and the pressure that is released shoots the ice—along with all the yeast sediment—out of the bottle. In times past, this opening process also was done by hand. A little wine is always lost during *dégorgement* and must be replaced. This brings us to the last and arguably most secretive step of champagne making: *dosage*.

Dosage

Dosage consists of adding a small amount of "liquor" (a mixture of sugar and old wines) to the bottle to replace the bit of wine lost during *dégorgement*. Each and every champagne house has a well-kept secret formula for this liquor. The wide variation in the levels of sweetness of a champagne—from *brut* (very dry) to *doux* (very sweet)—depends on the amount of sugar added at this point. Bottles are then quickly corked with traditional champagne corks and cages (to help keep the corks from popping out due to the pressure inside the bottles), receive labels, and finally are ready to drink!

○ ○ ○ ○

A FLUTE OR A GOBLET?

Now that we have a bottle of champagne all ready and waiting for us to taste, the question arises: How do we serve it? Red wine is best served in a glass with a full and open bowl. White wine traditionally is served in a glass with a more modest tulip shape. For champagne, what would be the most fitting kind of glass? Most people these days probably would choose a champagne flute, and there are good reasons to do so. However, the flute was not always the vessel of choice when it came to drinking champagne.

The champagne goblet, or *coupe*, was very fashionable in France from the early 1700s until the 1970s; legend has it that this glass was modeled on the celebrated breasts of Madame de

Pompadour, mistress of King Louis XV, who reigned from 1715 until 1774. Madame de Pompadour's appreciation of this special drink was well known; she was reputedly delighted with champagne and is said to have proudly explained to the court ladies who tried to guess the secret of her beauty and fresh appearance that it was due only to "the wine which allows you to look your best the morning after a tumultuous festivity." Legend also has it, though, that the *coupe* originally was modeled in porcelain on the breasts of the famed Marie Antoinette, the queen of Louis XVI, in the late eighteenth century. And still others say that the *coupe* had its origin in England, where it wasn't designed with anyone in mind but was made specifically to drink champagne. Suffice it to say that the origin of the *coupe* is still the source of some interesting cocktail party deliberations but remains uncertain.

Although it is still popular, the *coupe* actually does not allow the taster to fully enjoy the qualities of the champagne within it (although some *coupe* lovers may disagree). Shallow and wide-brimmed *coupe* glasses tend to be unstable and likely to spill. Furthermore, *coupes* do not present the wine's elegant bubble trains to best visual advantage. Nowadays, in formal settings champagne is served in a long-stemmed flute—a slender, elongated glass with a deep, tapered bowl. The tall, narrow configuration of the flute both highlights and extends the flow of bubbles rising to the crown, and the restricted open surface area concentrates the flavors carried up with the bubbles and released by their collapse. In contrast, the large open surface area of the *coupe* increases the discharge of gas, whereas the slender stemware of the flute prolongs the drink's chill and helps it to retain its effervescence.

However, before you uncork and pour your bottle of champagne, keep in mind the old saying: "The ear's gain is the palate's loss." Rather than popping the cork with a bang, instead, the cork should be released with a subdued sigh. A loud uncorking is the sign of a champagne-opening amateur. However, this is not to say that there isn't some fun in being deliberately amateur about uncorking a bottle of champagne. . . . Just keep in mind that an uncontrolled champagne cork popping out of a bottle can reach a velocity of about 50 kilometers per hour, so if it hits someone in the eye, it could do some serious harm and dramatically change the course of any romantic evening you might have had planned.

Another dramatic event that sometimes can occur on opening of a bottle of champagne is gushing, or *gerbage* as it is called in France. *Gerbage* is caused by excessive bubble production inside the bottle, which can produce a sudden foamy rush of champagne when the bottle is opened. Racing drivers actually like to intensify the phenomenon by vigorously shaking the bottle before opening it, which forces the gaseous carbon dioxide present in the headspace under the cork into the rest of the liquid. This causes many tiny bubbles to get trapped and float under the liquid surface. When the bottle is opened, the fall in pressure makes all these bubbles expand so quickly that they try to take up all the space in the bottle and, as a result, push the liquid champagne out of the bottle in an uncontrollable rush of foam. Though an episode of gushing is fun for some, champagne manufacturers actually try to prevent this from happening at their wineries. Since bottles on the production line knock against each other continuously—jostling each other's contents and forcing extra bubbles of carbon dioxide into the champagne—gushing sometimes happens during the *dégorgement* process before dosage. This can cause significant delays on the production line, not to mention loss of champagne, so champagne makers try to be quite careful to keep the speed of the production line down so that any jostling and knocking about are kept to a minimum.

○ ○ ○ ○ ○

THE BIRTH OF A BUBBLE

Why Does Champagne Make Bubbles?

The gas responsible for bubble production is carbon dioxide, which is produced by yeast during the second fermentation in the sealed bottle. According to Henry's law, equilibrium is established between carbon dioxide molecules dissolved in the liquid and carbon dioxide molecules in the vapor phase in the headspace under the cork. Before opening the bottle, the pressure of the carbon dioxide under the cork is about 6 atmospheres. The amount of dissolved carbon dioxide molecules in equilibrium is about 12 grams per liter of champagne. When the bottle is opened, the carbon dioxide pressure in the vapor phase suddenly drops, the thermodynamic equilibrium of the closed bottle is broken, and the liquid is supersaturated, which means that it now contains an excess of carbon dioxide molecules in comparison with the atmosphere outside the bottle. To recover a new and stable thermodynamic state corresponding to the partial pressure of carbon dioxide molecules in the atmosphere, almost all the carbon dioxide molecules dissolved into the champagne must escape.

If we assume that a classic champagne flute contains about 0.1 liter of champagne, we can estimate that approximately 0.7 liter of gaseous carbon dioxide must escape from it in order for equilibrium to be regained. To get an idea of how many bubbles this involves, we can divide the volume of the gaseous carbon dioxide in the flute (0.7 liter) by the volume of an average bubble (about 500 micrometers in diameter), and if we do so, we discover that a huge number of bubbles needs to escape a flute of champagne before the liquid can reach equilibrium: almost 11 million bubbles—more than the population of New York City!

When champagne is poured into a glass, there are two ways in which dissolved carbon dioxide molecules can escape from this supersaturated liquid: (1) directly through the champagne surface and (2) through bubble formation (Figure 2). Recent experiments have shown us that in a classic crystal flute, only about 20 percent of the carbon dioxide molecules escape in the form of bubbles, whereas the other 80 percent escape

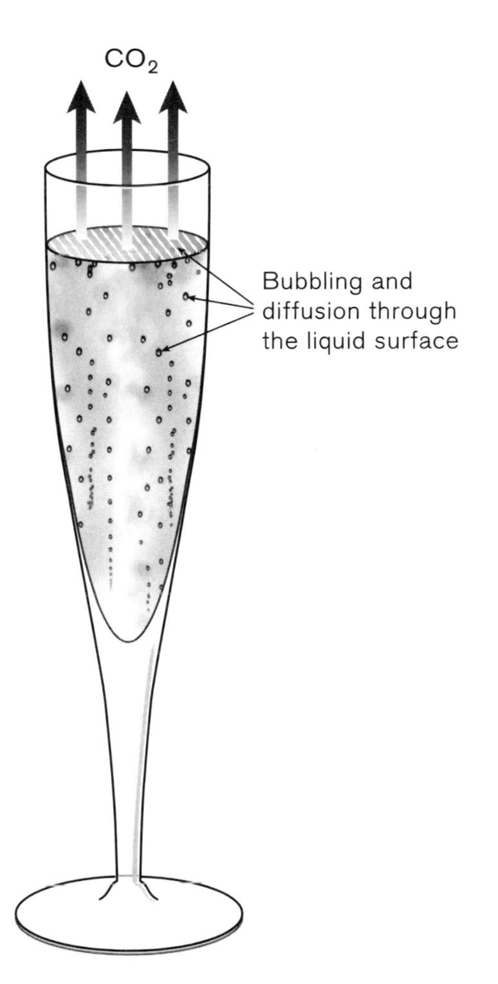

Figure 2 The two methods of carbon dioxide discharge in a flute of champagne.

directly through the free surface of the champagne. Thus, if you were to leave your glass of champagne alone and resist drinking from it until it stopped sparkling, about 2 million carbon dioxide bubbles would have escaped from your flute. But how do these bubbles form, or *nucleate*, in the first place?

Bubble Nurseries

In liquids supersaturated with gas molecules, bubbles don't just pop into existence *ex nihilo*, or from out of nothing. To form into bubbles, gas molecules in liquids such as champagne must cluster together and push their way up through the liquid molecules that are held together by Van der Waals attractive forces.* Therefore, bubble formation is limited by an energy barrier that is inherent to

*Van der Waals forces are relatively weak electrical forces that attract neutral (uncharged) molecules to each other in almost all organic liquids and solids.

the champagne. To get a better sense of what is happening here, imagine blowing up a balloon—it's difficult in the beginning when the balloon is small, but it becomes much easier as the balloon gets larger. The same principle holds true for bubble nucleation in champagne. The genesis of bubbles requires a huge amount of energy, but once bubbles are formed, they require less and less energy as they expand.

Bubble formation in carbonated beverages requires preexisting gas cavities of a certain size in order for the molecules of carbon dioxide to overcome the nucleation energy barrier and grow freely into bubbles. When a bottle of champagne is opened, classic nucleation theory tells us that the *critical radius* (or, simply, the size) below which bubble production is impossible is around 0.2 micrometer. To actually see these bubble production sites (or "bubble nurseries"), my team and I developed a workbench equipped with a high-speed video camera with a microscopic lens able to film up to 3000 frames per second with micrometric resolution (Figure 3). Sure enough, we saw that there are tiny preexisting gas cavities on the walls of the glass that are greater than this critical radius of 0.2 micrometer. This kind of bubble nucleation from preexisting gas cavities is referred to as *nonclassic heterogeneous nucleation*. In contrast, scientists talk about *homogeneous nucleation* when bubbles appear directly in the bulk of a liquid without the help of any preexisting bubble production sites.

Contrary to popular belief, however, bubble nucleation sites aren't found on scratches or irregularities in the glass itself; we know this because the width of glass or crystal faults is far below the critical radius of curvature required for bubble creation.* In fact, bubble nucleation sites are located on impurities that are stuck *on* the glass wall. Most nucleation sites are elongated, hollow,

*The critical radius of curvature required for bubble creation at the opening of a bottle of champagne is 0.2 micrometer.

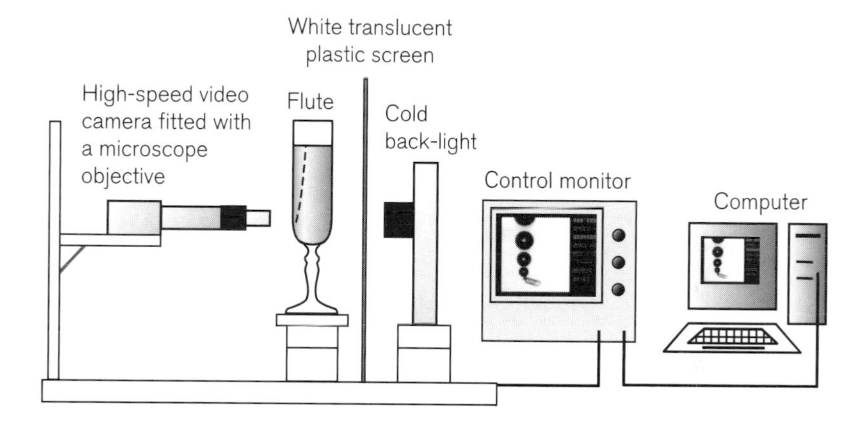

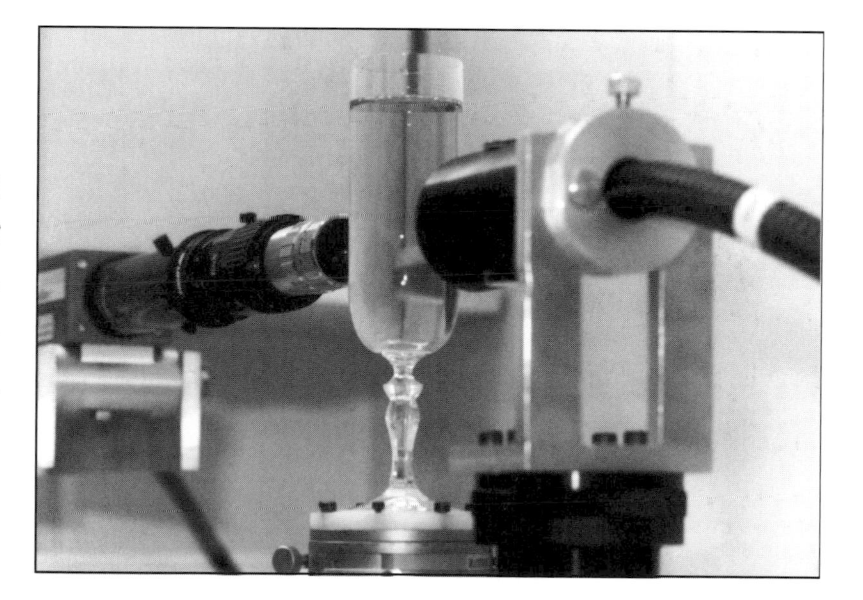

Figure 3 Scheme and photographic detail of the workbench used to visualize bubble nucleation sites. (Photograph by Gérard Hamalian.) © 2002 American Society for Enology and Viticulture.

roughly cylindrical cellulose fibers cast off from paper or cloth that floated onto the glass from the surrounding air or that remained on the glass when it was wiped dry after washing. Within each fiber is a cavity called a *lumen*. Because of their geometric properties, such hollow particles cannot be wetted down completely by the champagne when it is poured into the glass, and they consequently trap tiny pockets of air when the flute is filled.

It's strange to think that champagne bubbles spring to life from "dirt" or small particles of debris contaminating the surface of the glass. However, imagine a perfectly clean flute wall without any adherent particles whatsoever. Champagne poured into such a "clean" container would not bubble at all. All the excess carbon dioxide molecules would escape directly through the free surface of the liquid. Such an experiment was conducted in Möet & Chandon's laboratory, and sure enough, after pouring, the champagne looked simply like a still wine.

Incidentally, you also may have noticed and wondered why very few bubbles nucleate in a freshly opened bottle of champagne before pouring. This phenomenon is not due to a lack of particles originally present on the glass of the bottle. Rather, it is because the long storage over several years has succeeded in wetting nearly every available gas pocket inside the bottle, thus completely preventing the dissolved carbon dioxide molecules from escaping in the form of bubbles. However, when the champagne is poured into a flute, you'll notice that the bubbling and sparkling are plentiful.

Eight typical nucleation sites found in a glass of champagne are shown in Figure 4. The gas pockets trapped inside the particles are very clearly visible. These gas pockets are larger than the critical radius that we know is necessary to breach somehow in order for bubbles to form in the liquid, and subsequently, we know that dissolved carbon dioxide molecules migrate from within the liquid of the champagne into the lumens of the fibers. The gas pockets

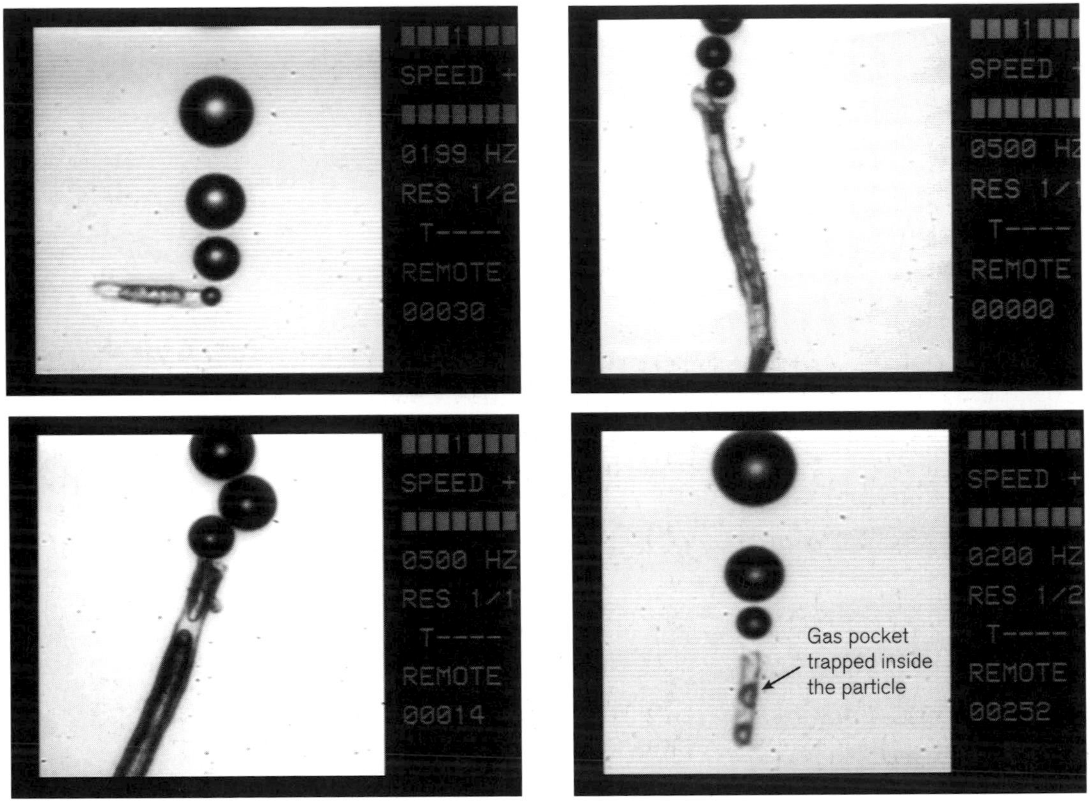
SPEED +
0199 HZ
RES 1/2
T----
REMOTE
00030
SPEED +
0500 HZ
RES 1/1
T----
REMOTE
00000
SPEED +
0500 HZ
RES 1/1
T----
REMOTE
00014
SPEED +
0200 HZ
RES 1/2
T----
REMOTE
00252
Gas pocket trapped inside the particle

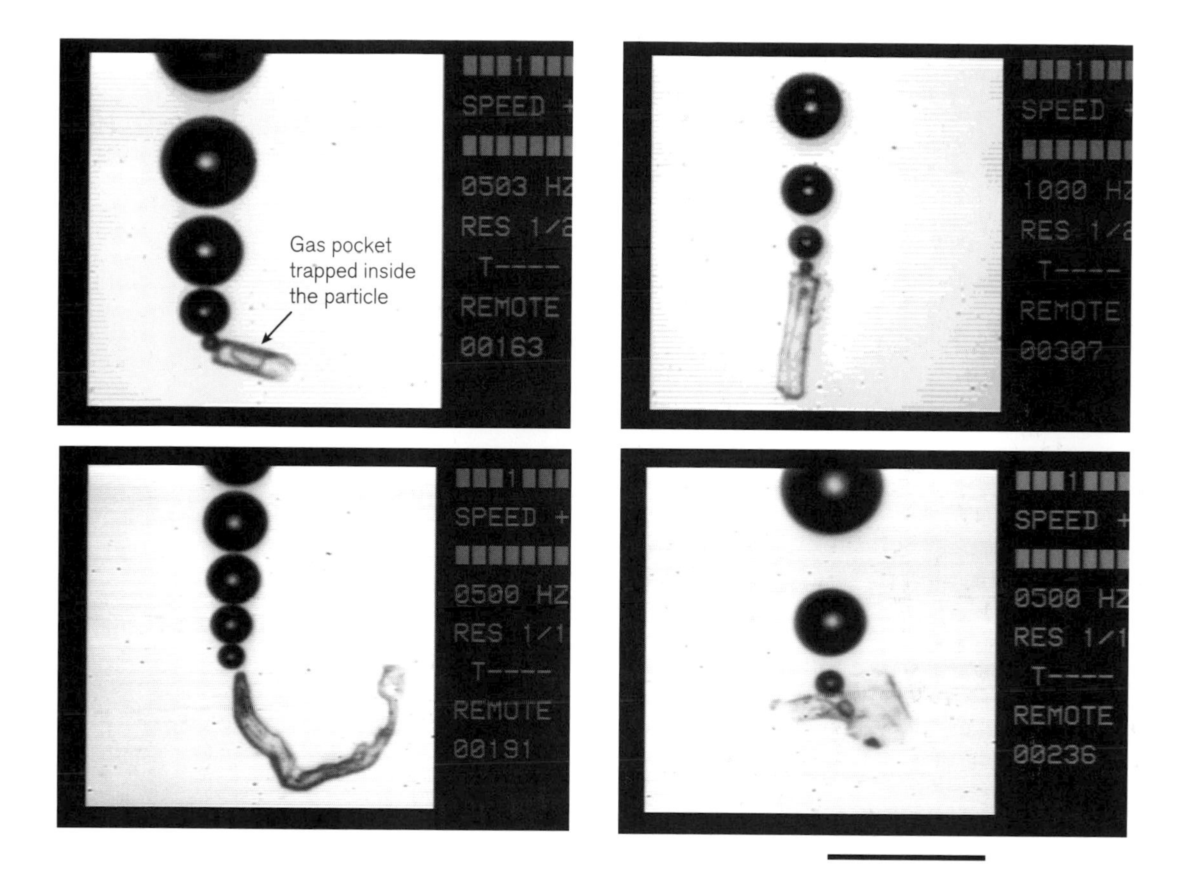

Figure 4 Micrographs of eight particles acting as nucleation sites in a flute of champagne. The gas pockets trapped inside the particles can be seen clearly (bar = 100 micrometers). © 2002 American Society for Enology and Viticulture.

trapped inside the fibers grow in size with the continual accumulation of carbon dioxide molecules, and finally, a bubble is ejected from the end of the fiber and sometimes even from both ends (Figures 5 and 6).

A Repetitive and Clockwork Bubbling Process

A complete time sequence showing the cycle of bubble production from its beginnings in a cellulose fiber is displayed in Figure 7. Between frames 1 and 4 (approximately 30 milliseconds), a bubble grows, rooted to its nucleation site because of capillary forces—that is, because a contact exists between the bubble and the fiber. Actually, a bubble growing rooted to its nucleation site is very like a drop of water growing at the edge of a faucet just before gravity forces the drop to detach and fall. It is the increasing buoyancy of the bubble as it expands that finally causes it to detach (in frame 5). However, a small gas pocket still remains trapped inside the

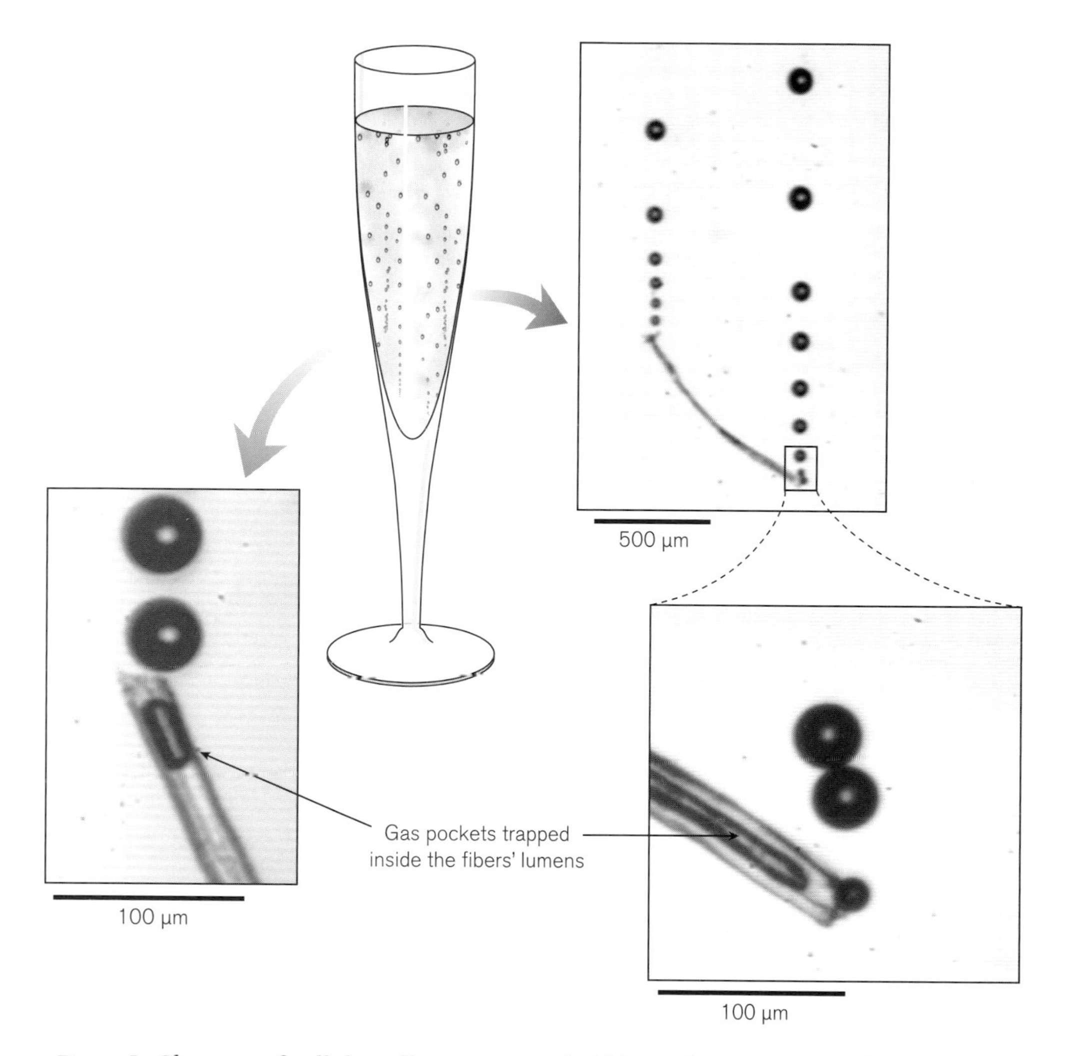

Figure 5 Close-up of cellulose fibers acting as bubble nucleation sites on the wall of a flute of champagne as seen through the microscope objective of a high-speed video camera. The gas pockets trapped inside the lumens of the fibers and responsible for the repetitive production of bubbles are seen clearly as dark areas.

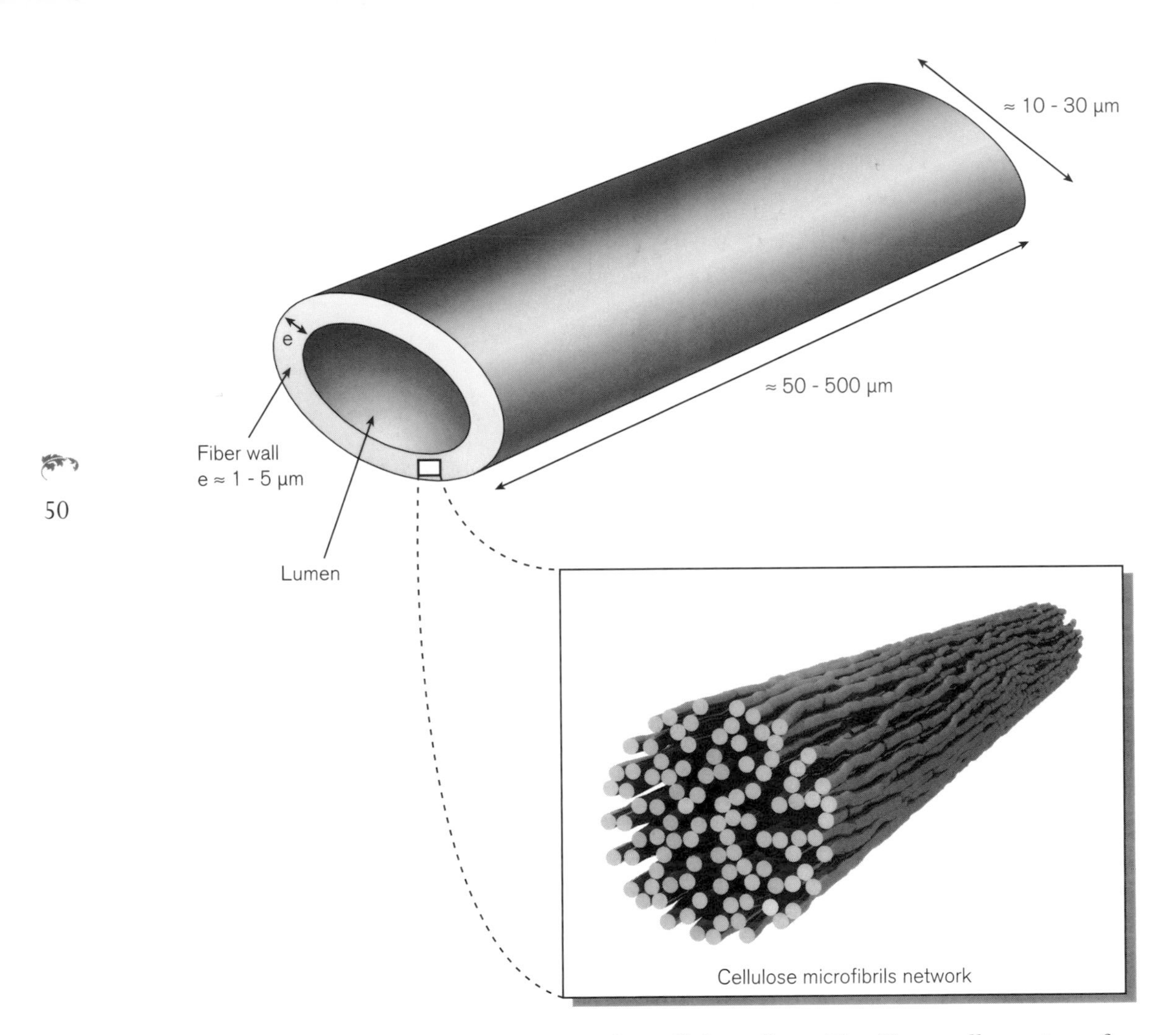

Figure 6 Schematic representation of a cellulose fiber. The fiber wall consists of closely packed cellulose microfibrils oriented mainly in the direction of the fiber. (Drawing courtesy of Daniel Topgaard.)

fiber's lumen, and this is the environment that allows new bubbles to nucleate, grow from their nucleation sites, and detach. A new bubble appears in frame 8, and it will meet exactly the same fate as the previous one—and so on until production finally stops due to a lack of remaining dissolved carbon dioxide molecules in the champagne. When released from its nucleation site, a bubble corresponds to the size of the fiber that spawned it, which generally is about 10 to 20 micrometers in diameter.

Preexisting gas cavities trapped inside particles stuck on the glass wall are what we see in the photographs as tiny "bubble guns" that release streams of bubbles up through the champagne. The cycle of bubble production at a given nucleation site has a *bubbling frequency*, which is the number of bubbles produced per second. This clockwork bubble production is seen easily with the naked eye by lighting up bubble trains with a stroboscope and matching the flash frequency of the strobe light to the frequency of the

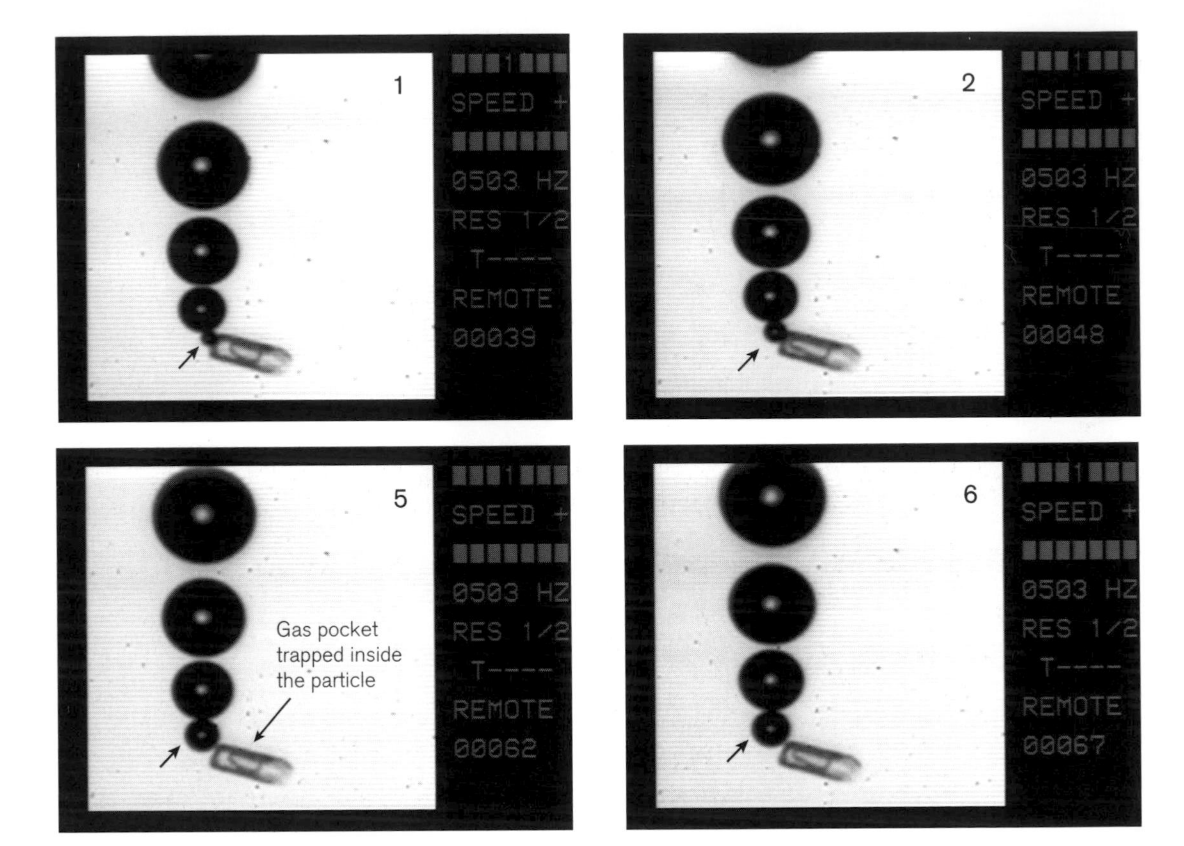
1
SPEED +
0503 HZ
RES 1/2
T----
REMOTE
00039
2
SPEED +
0503 HZ
RES 1/2
T----
REMOTE
00048
5
SPEED +
0503 HZ
RES 1/2
T----
REMOTE
00062
Gas pocket trapped inside the particle
6
SPEED +
0503 HZ
RES 1/2
T----
REMOTE
00067

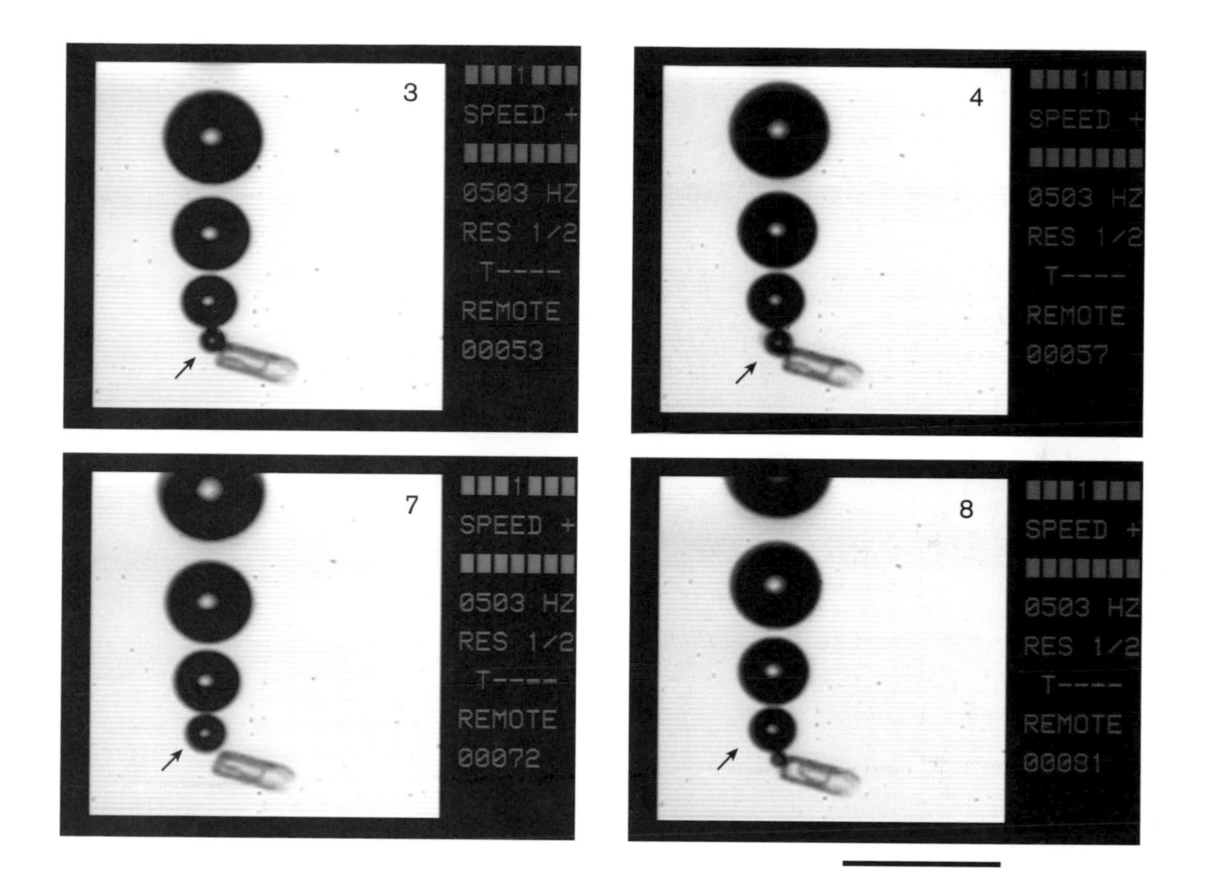

Figure 7 Typical time sequence illustrating the bubble nucleation process. The time interval between any two frames is approximately 10 milliseconds. Black arrows point to the initial bubble (bar = 100 micrometers). © 2002 American Society for Enology and Viticulture.

bubble production cycle. Between two flashes of light, a bubble in the train takes the place of the bubble just preceding it, and the corresponding bubble train appears "frozen." Because the kinetics of bubble formation depend on the content of dissolved carbon dioxide, the bubble formation frequency of a given nucleation site decreases progressively as time passes and the carbon dioxide content is diminished gradually until ebullition eventually stops.*

The kinetics of bubble production also depend on the size and shape of the particles that act as nucleation sites. In a champagne-filled flute, the collection of particles on the flute wall most likely will be made up of all shapes and sizes. Therefore, bubble trains with various bubble formation frequencies may be observed in the

* Ebullition actually stops before the partial pressure of carbon dioxide in the champagne gets down to 1 atmosphere, that is, before the champagne reaches equilibrium with the atmosphere. If the wine is then left to rest and not drunk, the remaining carbon dioxide in the champagne is lost into the air through *volatilization*, or the passing off of carbon dioxide from the wine in the form of vapor.

same flute at the same time (Figure 8). Moreover, since the kinetics of bubble formation also hinge on the content of dissolved carbon dioxide in a liquid, bubble formation frequencies actually vary from one carbonated beverage to the next. For example, in Champagne wines, the dissolved gas content is approximately three times higher than in beer, and the most active nucleation sites emit up to about thirty bubbles per second; in beer, bubble nurseries emit only up to about ten bubbles per second.

Classic nucleation theory also tells us that the critical radius required to enable bubble formation from preexisting gas cavities is not fixed. It actually changes over time within a glass of champagne and is inversely proportional to the dissolved carbon dioxide content of the champagne. As a result of bubbling and diffusion through the surface of the champagne, carbon dioxide molecules escape progressively from the liquid. Subsequently, the dissolved carbon dioxide content in the liquid decreases progressively, and

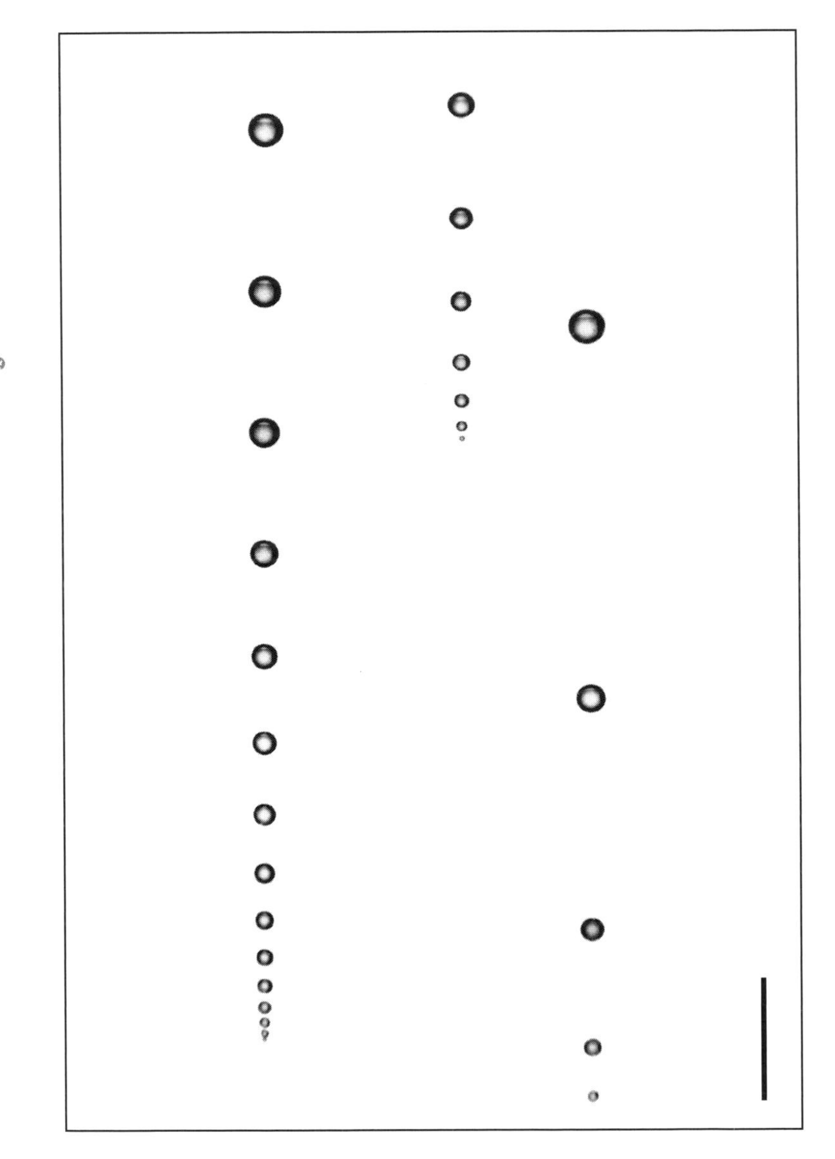

Figure 8 In the same flute and at the same time, because there is a collection of particles of various shapes and sizes, bubble trains of various formation frequencies may be observed (bar = 1 millimeter). (Photograph by Gérard Liger-Belair.)

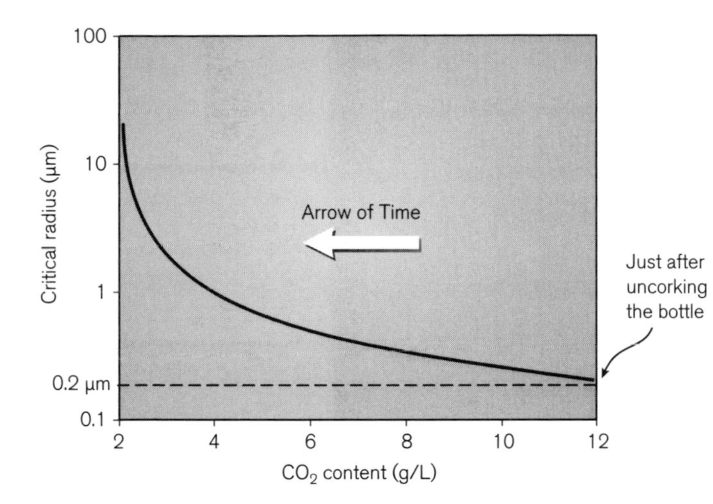

Figure 9 During the process of carbon dioxide discharge, the critical radius below which bubble production becomes thermodynamically impossible increases progressively. As a result, nucleation sites, in turn, become increasingly inactive in direct proportion to the size of the embryonic bubbles trapped inside the particles.

as a result, the corresponding critical radius required for bubble formation *increases* (Figure 9). As time continues to move on after pouring the champagne into the flute and carbon dioxide levels in the champagne decrease, bubbles originating from the smallest particles eventually are unable to breach the critical radius for bubble production, and these nucleation sites cease producing bubble trains. Eventually, bubble production stops in all the particles acting as nucleation sites, with the largest sites ceasing production last.

○○○○○○

THE BUBBLE RISES

Growing Bubbles

After bubbles are born and released from their nucleation sites, they rise toward the liquid surface in elegant bubble trains and grow in size during their ascent (Figure 10). After a journey of about 10 centimeters, champagne bubbles will reach a diameter close to 1 millimeter—meaning that they increase by a factor about 1 million in volume during their trip up (since their size at the point they leave their nucleation sites was about 10 micrometers).

So what actually makes bubbles grow as they rise toward the liquid surface? In fact, bubble growth is due to excess carbon

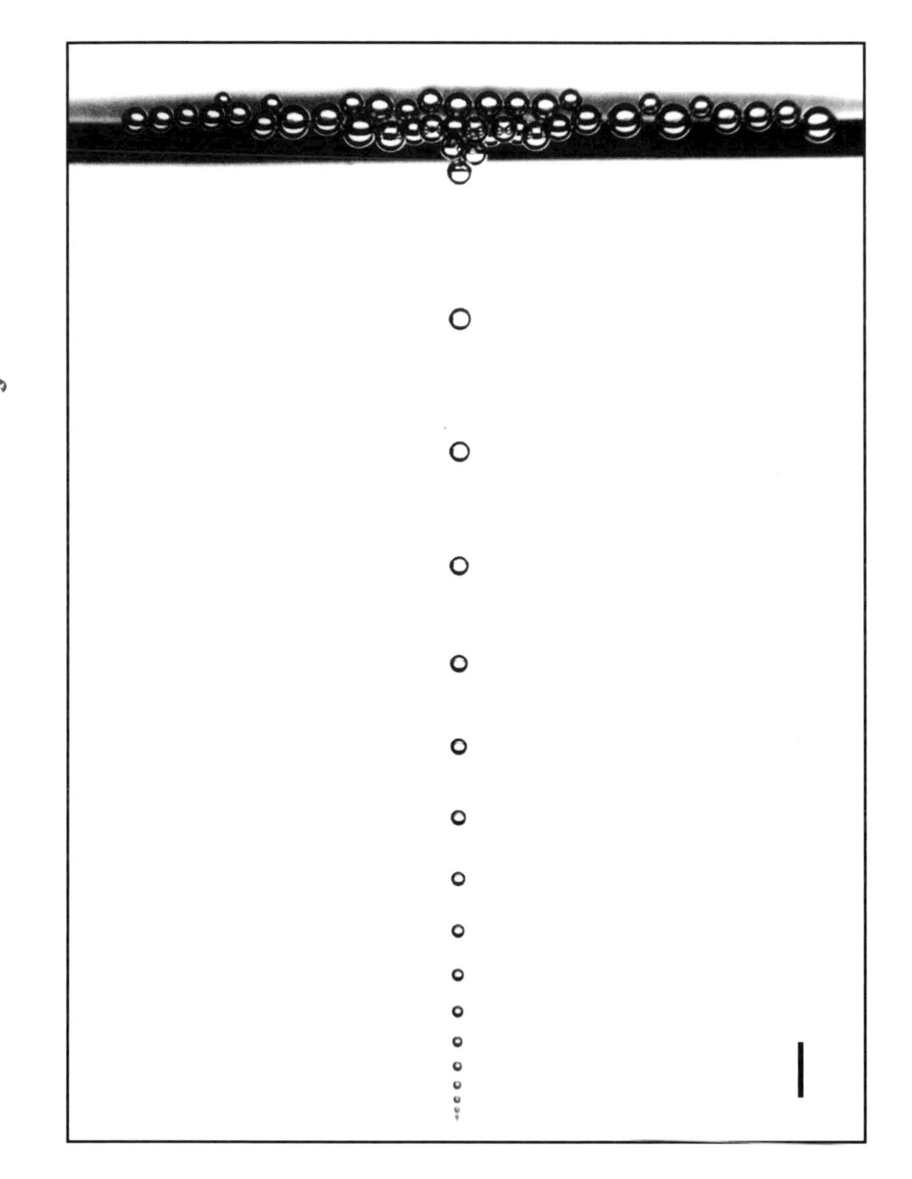

dioxide molecules that continue to migrate into the bubbles as they rise because the bubbles ejected from their nucleation sites already have a size larger than the critical radius required for them to grow freely. Therefore, the huge bubble growth observed during ascent is caused by a continuous diffusion of dissolved carbon dioxide through the bubble interface into the bubble itself.

As far as Champagne wines are concerned, the bubble growth rate is very important because this rate determines the final average size of the bubbles—and this, in turn, is a hallmark of the wine's quality. One source of influence on the bubble growth rate is the overall dissolved carbon dioxide content of the champagne. Since, as we know, carbon dioxide escapes progressively from the champagne when it is poured into a flute (directly through the liquid surface

Figure 10 (Left) A typical bubble train. The dark line running horizontally across the top of the picture is due to the liquid meniscus between the free surface and the glass wall (bar = 1 millimeter). (Photograph by Gérard Liger-Belair.)

and through the bubbling process), the bubble growth rate of ascending bubbles decreases progressively over time—as you probably have noticed yourself while enjoying a glass of champagne.

You also may have noticed that old champagnes and sparkling wines tend to give a greater number of little bubbles than young champagnes and sparkling wines do. Why? The answer is quite simple. Champagne corks and cages are not completely hermetic, and consequently, minute traces of carbon dioxide escape slowly during the aging process in the cellar. When the bottle is finally opened several years later, the dissolved carbon dioxide content is less than it was *before* the aging process, thus decreasing bubble growth in comparison with that of a young champagne. This is probably the origin of the saying, "The smaller the bubbles, the better the wine." Aging is certainly needed to develop the so-called roundness and fine flavor of champagne, but a well-aged champagne is recognized on tasting not only by the general complexity of its flavor but also by the complementary fineness of its effervescence.

Accelerating Bubbles

We now know how bubbles are formed and that they grow when they ascend through the champagne within a glass, but what causes a bubble to rise at all? Buoyancy (also known as *Archimedes' principle*) is obviously the driving force behind the ascent of champagne bubbles, but a bubble also encounters resistance from the liquid molecules surrounding it, which drag against the bubble and retard its ascent. Because of their expansion during their ascent, bubbles further increase their buoyancy, which is a function of (or directly proportional to) the bubble volume. The drag force exerted on expanding bubbles is insufficient to inhibit their increasing buoyancy, so bubbles accelerate as they travel up through the champagne and separate gracefully from each other as they approach the surface (see Figure 10).

Some of you may be wondering, given that gas bubbles need gravity to rise (by way of buoyancy), what champagne or beer would be like in the absence of gravity—say, on a space station, for example. After uncorking a bottle in a place with zero gravity, bubbles nucleated on the bottle wall would be completely unable to rise. Carbon dioxide bubbles would form and grow, but they would remain rooted to and would not be able to detach from their nucleation sites. The bubbles would grow bigger and bigger inside the bottle and would very quickly replace the liquid—which would then overflow out of the bottle.

With this not-so-ideal scenario in mind, we also could go on to ask ourselves, Where in the solar system would we be able to enjoy the tiniest, finest bubbles in a flute of champagne—on the Moon or Jupiter? Since bubbles grow in size during their way up to the liquid surface, the faster the velocity of the bubble as it rises, the less time it takes to travel up through the champagne to the surface, the less chance it has to grow—and the smaller the bubble will be. Since gravity is the driving force behind bubble rise, the velocity of the bubbles in a flute of champagne is directly related to the influence of gravity wherever you are. On Earth, the acceleration of gravity is 9.8 meters per second squared. On Jupiter, the most massive planet in our solar system, gravity acceleration is about a hundred times that of Earth, so it follows that in our solar system the best place to get the tiniest bubbles in your flute of champagne would be on the most massive planet, where the gravity is strongest.

A Surfactant Shield

Champagne, sparkling wines, and beers are not pure liquids. In addition to water, they contain alcohol, dissolved carbon dioxide molecules, and many other organic compounds. The molecules of these organic compounds have both water-soluble and water-insoluble parts (Figure 11). Such substances are known as *surfactants* (short for "surface-active agent"). Surfactants are attracted to the surfaces of bubbles and position their water-insoluble parts inside the gas-filled

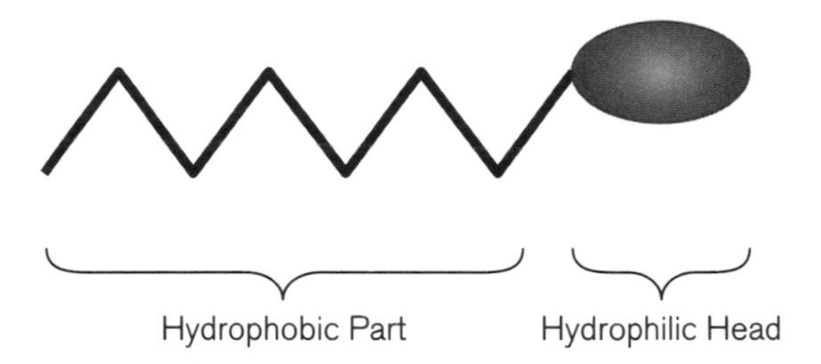

Figure 11 Schematic representation of a surfactant molecule.

bubble by a process called *adsorption* so that they do not touch the liquid surrounding the bubble. Eventually, surfactants form a kind of coat around a bubble. The role of this surfactant coating around gas bubbles becomes very important when buoyancy causes bubbles to detach from the glass wall of the flute and plough their way through the liquid molecules. Adsorbed surfactants stiffen a bubble by forming a sort of shield—a rigid wall—on its surface (Figure 12). According to fluid dynamics theory, a rising bubble made rigid by surfactants runs into more resistance than a bubble that maintains its original flexible skin free from surface-active materi-

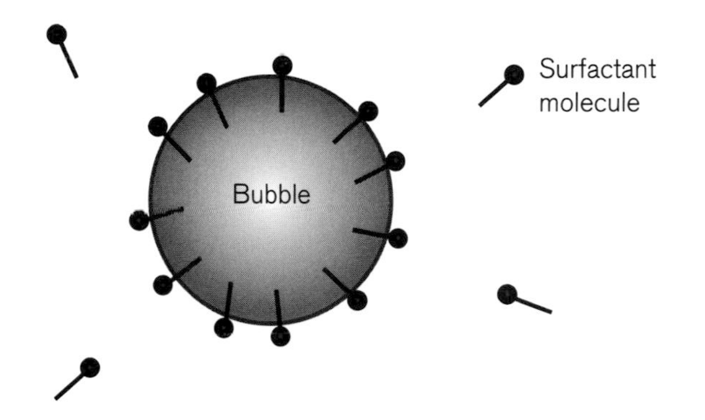

Figure 12 Bubbles act as traps for surfactants.

als. The more surfactants there are, the more drag is exerted on a bubble rising in a surfactant-filled solution; this drag only continues to increase, and by the time the bubble interface is completely contaminated, the bubble's velocity is decreased dramatically. For example, in ultrapure water, that is, water free from all surfactants, a millimetric bubble rises at a velocity close to 30 centimeters per second. In water with only minute traces of surfactants (on the order of several milligrams per liter), bubble velocity decreases as

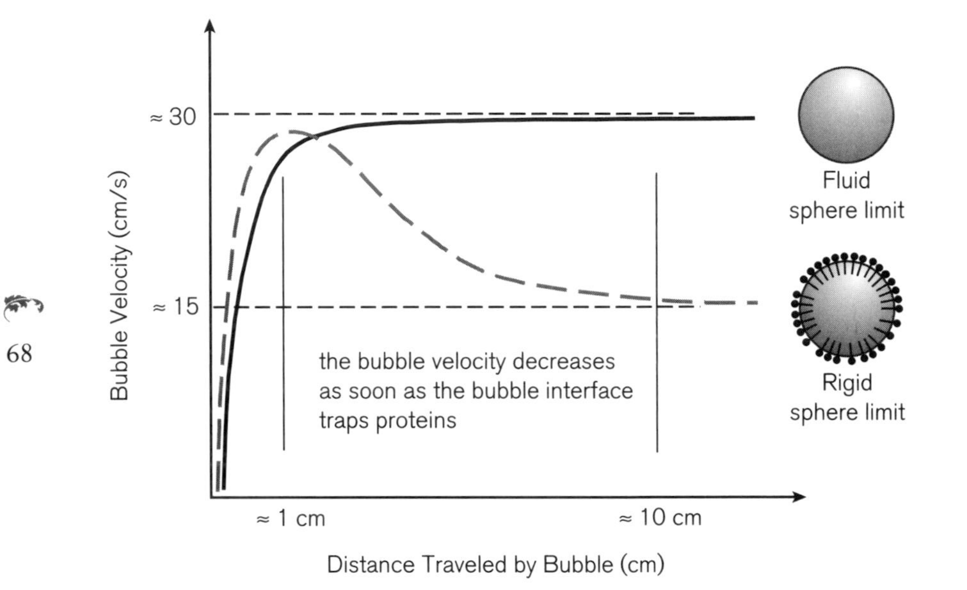

Figure 13 Velocity (in centimeters per second) versus distance traveled (in centimeters) by a millimetric bubble released in ultra-pure water (black line) and in water to which has been added 10 milligrams per liter of protein (dashed line). Redrawn with permission from Ybert (1998).

soon as the bubble interface traps surfactants, and by the time the bubble is completely rigidified by a surfactant coating, it is traveling at a velocity of about 15 centimeters per second (half the velocity of a bubble rising through pure water) (Figure 13).

Most of the time bubbles rising in surfactant solutions are only partly rigidified. This is so because the flow of liquid around the rising bubble acts to sweep the surfactants around toward the bottom of the bubble, and if there is not enough surfactant to keep the top part rigid, the bubble will have a clean upper part and a rigid lower part (Figure 14). Its velocity then will be between the fluid-sphere and rigid-sphere limits shown in Figure 13.

Self-Cleaning Bubbles

However, since a champagne bubble expands continuously as it rises through the supersaturated liquid that is champagne, the bubble interface increases continuously and therefore continues to offer

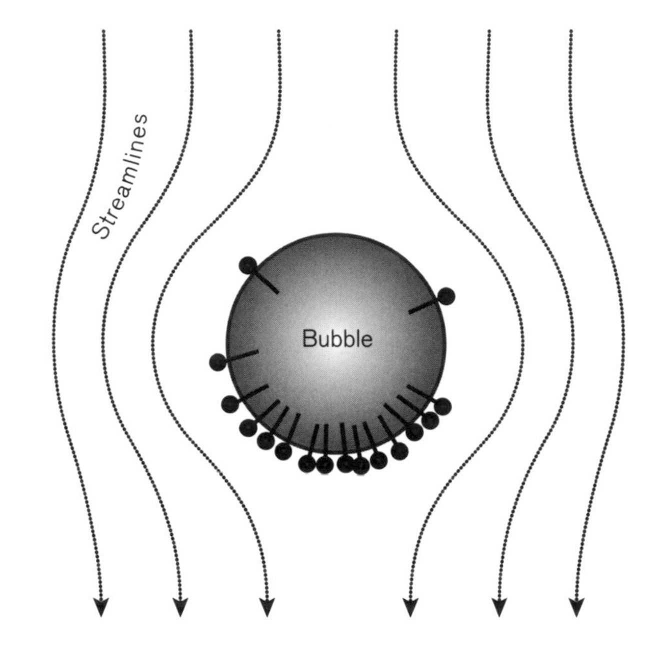

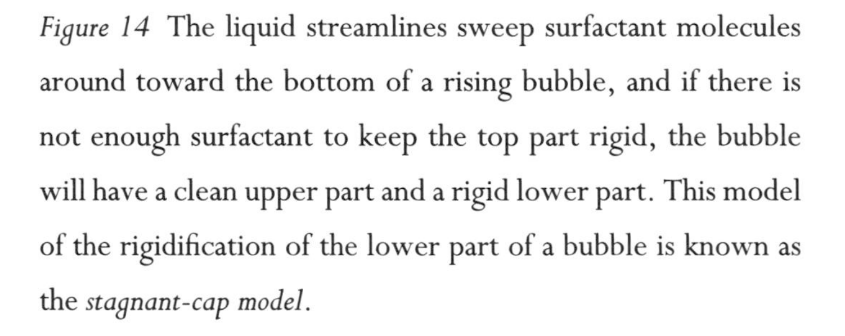

Figure 14 The liquid streamlines sweep surfactant molecules around toward the bottom of a rising bubble, and if there is not enough surfactant to keep the top part rigid, the bubble will have a clean upper part and a rigid lower part. This model of the rigidification of the lower part of a bubble is known as the *stagnant-cap model*.

newly created surface area to the surfactants. Expanding bubbles consequently experience two opposing effects. If the bubble's rate of inflation is greater than the rate at which surfactants are able to stiffen the bubble surface, a bubble is able to "clean its surface" of surfactants. This means that the ratio of the bubble surface covered by surfactants to the surface *not* covered by surfactants decreases. On the other hand, if this ratio were to *increase*, the bubble surface would become contaminated entirely by a surfactant coating and would become rigid (Figure 15).

A Comparison between Champagne and Beer Bubbles

Before conducting my own experiments on the expansion of champagne and beer bubbles, I read a work on beer bubble ascent published in *Physics Today* (October 1991). Two chemists at Stanford

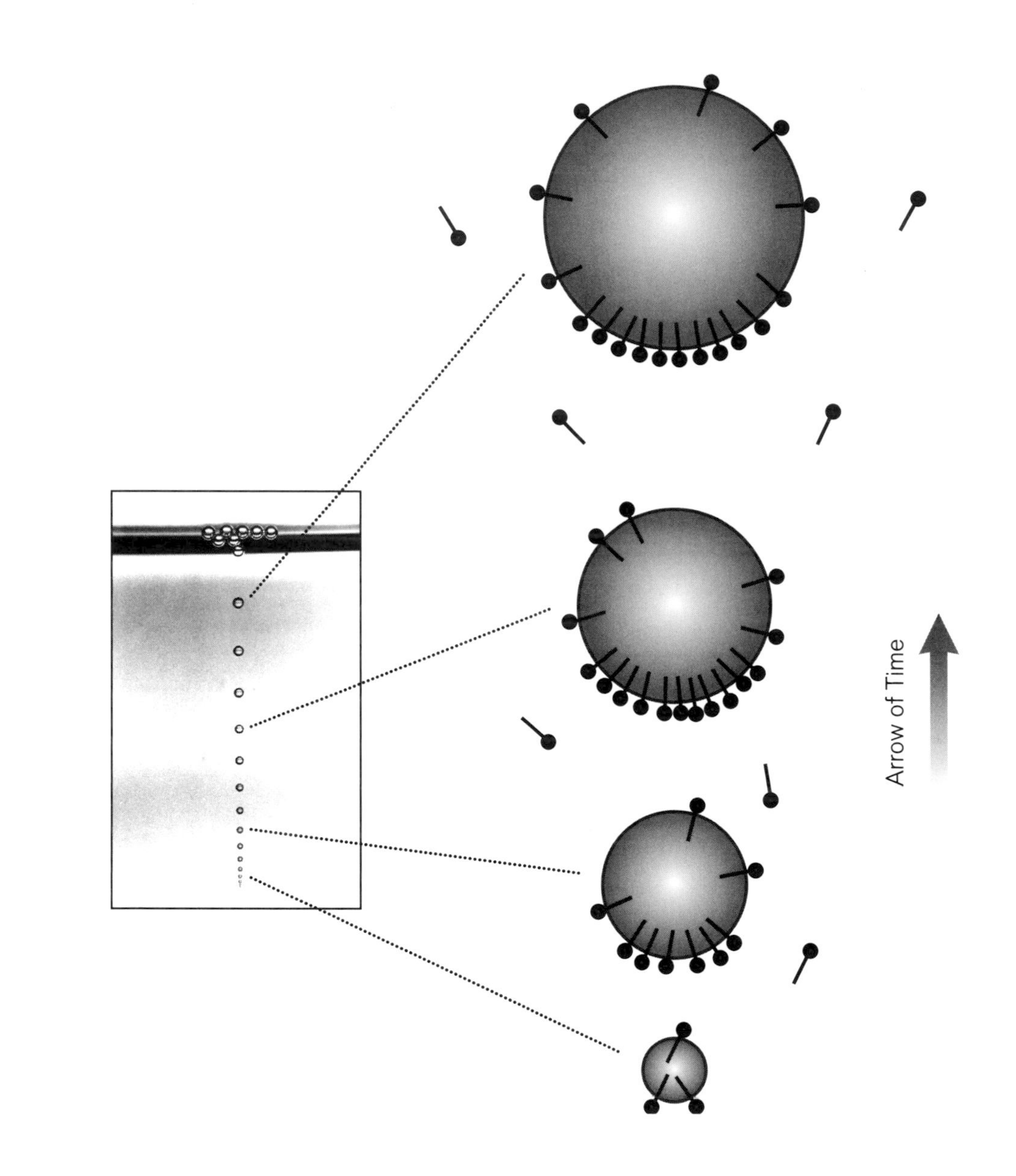
Arrow of Time

University, Neil Shafer and Richard Zare, reported that the motion of beer bubbles was similar to that of rigid spheres—meaning that the walls of beer bubbles probably were completely rigidified by surfactants. Since the chemical composition of champagne is rather different from that of beer, it wasn't clear that the results of Shafer and Zare would apply to bubbles in champagne and sparkling wines. If you look carefully at both a glass of champagne and a glass of beer, you see that the bubbles in the beer ascend at a more leisurely pace. Why is this?

The viscosity of beer cannot account for this because it is very close to the viscosity of champagne. However, by measuring the drag experienced by champagne bubbles versus beer bubbles and comparing these data with data on bubbles of various sizes ascending through all sort of different liquids, I discovered that, in fact,

Figure 15 (Left) As a bubble rises, there is a competition between surfactant adsorption and bubble growth.

beer bubbles did become rigid during ascent. When I examined bubble trains in champagne and sparkling wines, though, they told a different story. Champagne bubbles ascending toward the liquid surface changed from rigid spheres to fluid ones; as they expanded, the bubbles cleansed themselves of surfactants, experienced less drag as they rose through the champagne to the surface, and thus rose faster in comparison with the bubbles in beer (Figure 16).

This is actually not too surprising because two main chemical differences distinguish champagne from beer: First, beer contains a much greater quantity of surfactants (on order of several hundred milligrams per liter) likely to be adsorbed at a bubble interface than champagne (which contains only several milligrams per liter). Beer has many more surfactants than champagne simply because barley naturally releases more surfactants (mostly proteins and glycoproteins) than grapes do. Second, the dissolved gas content is approximately three times greater in champagne than in beer.

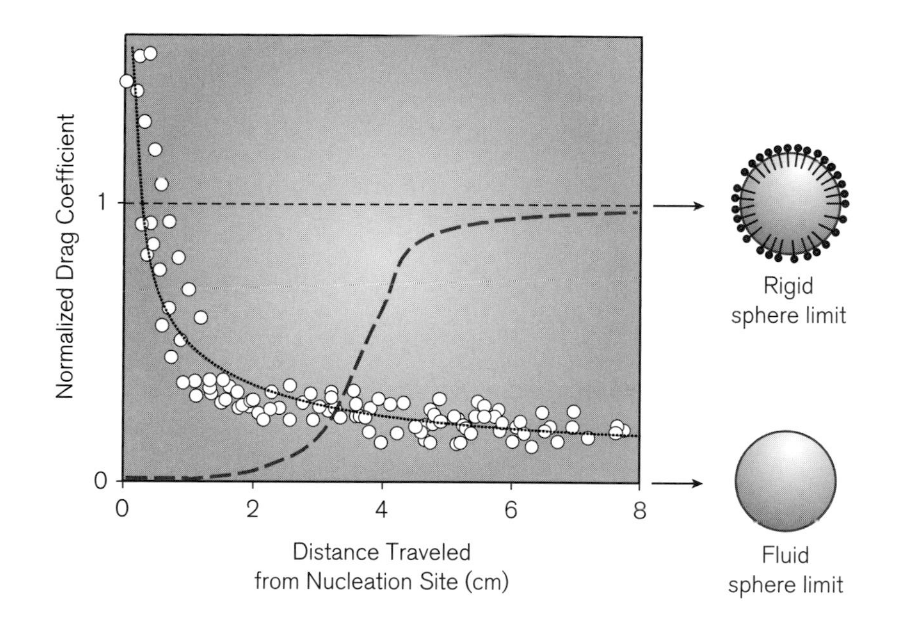

Figure 16 During their rise toward the surface, the bubbles in a flute of champagne change progressively from rigid spheres to fluid spheres, suggesting that they "clean themselves up" as they grow in size. Experimental data (○); fitted curve for experimental data (dotted line); behavior of a bubble of fixed size (dashed line).

Bubble growth rates, therefore, are correspondingly about three times higher in champagne. As a result, the "cleansing" effect in beer due to bubble expansion may be too weak to prevent rigidification of the beer bubble interface. In contrast, champagne bubbles expand too quickly (and there are simply not enough surface-active materials in the wine) for their surfaces to become rigid.

When Bubbles Go Crazy

If liquids such as champagne and beer were free from surfactants, the bubbles within them would not be able to rise up in their characteristically linear trains. In a pure liquid, bubbles rising in a line toward the surface are very unstable and typically do not maintain a straight line as they rise to the surface. The in-line instabilities of the bubbles are the result of interactions between successive bubbles rising up in a line. These bubbles are very vulnerable to any lateral displacement resulting from forces in the liquid surrounding them. If a bubble happens to be jostled a bit out of its line for any reason, the bubble actually will be lifted and pushed even further out of line by the next bubble coming up behind it, thus breaking the straight paths of the bubble trains and making the bubbles veer off in various directions as they rise.

In contrast, bubbles rising in liquids such as beer and champagne are stabilized by their surfactant coatings and thus can maintain an orderly line as they make their way up to the surface. If the rear part of a bubble is even partially rigidified by surfactants and this bubble runs into any force that displaces it away from its rising axis, the lift force it experiences actually tends to repel the bubble back *toward* the bubble train and its axis of symmetry rather than away from it, thus preserving the vertical stability of the line.

The stability of the bubble train also is influenced by the size of the bubbles in the train. Small bubbles (like those in champagne and beer) are spherical and rise in a straight line, partly due to the phenomenon just described. However, this would not be the case if the bubbles could grow to a critical diameter of roughly 3 millimeters. Bubbles of this size, no matter what the surfactant content of the liquid is, ascend at a higher velocity and become deformed by the pressure of the liquid pushing against them as they rise; the pressure of the liquid on the top parts of the rising bubbles deforms them in such a way that they appear rather ellipsoidal or egg-shaped. As a bubble swells in size, the deformation also continues to increase so that when the bubble's horizontal diameter reaches about twice its vertical diameter, it no longer rises in a straight line but rather follows a zigzag or spiraling path. You can see this phenomenon in fish tanks, where bubbles reach the critical size of about 3 millimeters and subsequently spiral up to the surface of the tank when they rise rather than floating up in orderly lines.

The journey of the bubbles in a glass of champagne or beer is actually too short (usually less than 10 centimeters) to let the bubbles expand to the critical size necessary to become deformed or potentially unstable. Within a *bottle* of champagne or beer, however, bubbles that nucleate on the bottom do reach this critical size, and they begin to oscillate. Such oscillations are believed to be a consequence of the wake instabilities* behind the rising bubble.

The photograph in Figure 17 illustrates this hydrodynamic instability. I took this picture immediately after opening a plastic bottle of clear soda. In the soda bottle, bubbles nucleated from the bottom of the bottle on defects in the plastic molding that trapped tiny gas pockets when the soda was poured. This time my goal was not to freeze the bubble motion but to get an idea of bubble trajectories. Therefore, I used a long exposure time: 1 second. The trajectories of the bubbles appear as white filaments because the bubbles reflect the ambient light as they rise up through the liquid. During their first 10 centimeters of ascent, bubbles rise in a straight

*Wake instabilities are caused by vortices that are formed behind rising bubbles when they reach a certain size—about 3 millimeters in diameter.

line; but after that, we begin to see the bubbles oscillate, and the amplitude of this oscillation grows as the bubbles swell in size and rise up through the bottle.

Pairing Off, Bouncing, and Coalescence

If the stability of bubbles rising in line seems to be assured by their surfactant shield, the vertical stability nevertheless still may be broken for the following additional reason: Bubbles released at high frequencies from their nucleation sites necessarily rise close to each other. Because the succes-

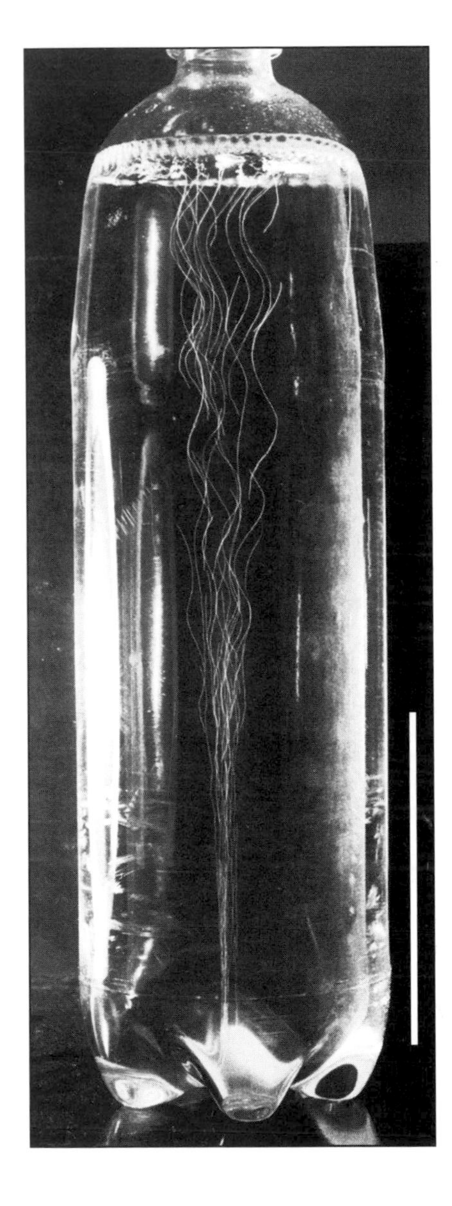

Figure 17 Bubble trajectories in a bottle of soda (white bar = 10 centimeters). (Photograph by Gérard Liger-Belair and Cédric Voisin.)

sive bubbles rise up through the liquid so close to one another, they begin to interact with each other hydrodynamically. For example, sometimes in a bubble train a trailing bubble may be pulled into the wake of a leading bubble. In the wake of a rising bubble, the pressure is slightly less than within the fluid around the rest of the bubble. Consequently, when a trailing bubble reaches the leading bubble's wake, it accelerates progressively through the wake, catches up with the leading bubble, and then finally bumps into it. The bounce of this impact causes the two bubbles to rotate rapidly into a cross-stream position, and they then continue to rise side by side.

Fluid physicists actually called this scenario, the *drafting, kissing, and tumbling scenario.* You already may know the term *drafting* or, at least, the phenomenon from watching cyclists riding close behind one another in a race. Drafting in this case is the act of riding behind another cyclist in an area of reduced air pressure that is created by the wake of the leading cyclist. In the drafting zone,

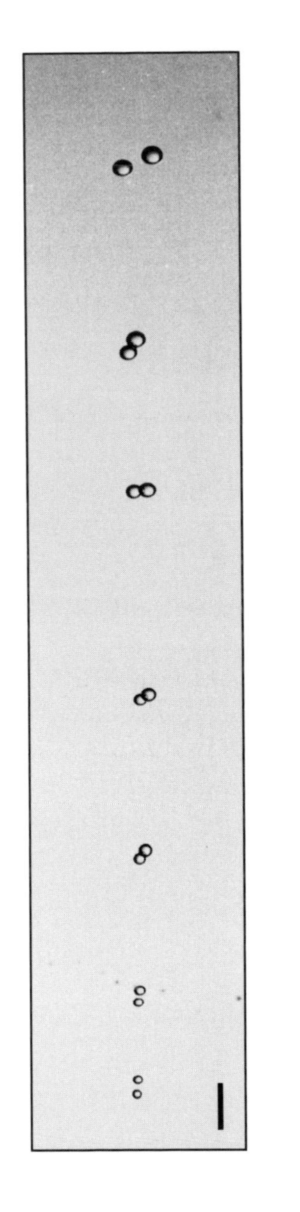

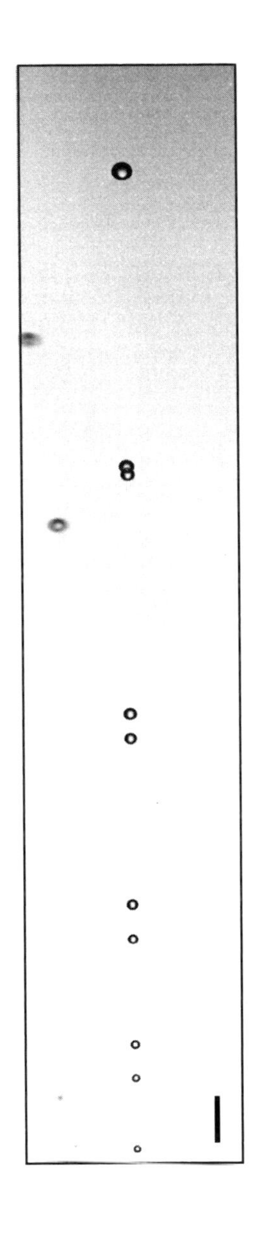

Figure 18 Two photographs illustrating (left) the drafting, "kissing," and tumbling process and (right) the drafting, "kissing," and coalescing process of the bubbles in a flute of champagne (bar = 1 millimeter). (Photographs by Gérard Liger-Belair.)

the trailing cyclist uses less energy to maintain the same speed as the leading cyclist. Similarly, in a liquid, the trailing bubble is able to move more quickly through the leading bubble's wake than the leading bubble is able to plow through the liquid above it.

In addition, after the *kissing* phase, bubbles don't always tumble; sometimes they even collide and actually combine with each other. However, this scenario depends on the velocity at which the bubbles hit, the respective sizes of the bubbles, and the sizes of their respective surfactant shields. The photographs displayed in Figure 18 illustrate both scenarios in a flute of champagne. Finally, because high bubble formation frequencies are needed to cause any interactions between successive bubbles at all, bubble pairings such as these therefore are best observed just after pouring.

○○○○○○○

THE BUBBLE BURSTS

The last step in a champagne bubble's life—and certainly the most spectacular—is its bursting at the liquid surface.

A Phenomenological Study

Bubbles begin to reach the free surface only a few seconds* after being born on an impurity below the liquid surface on the wall of the glass. At the liquid surface, the shape of the bubble results

* Depending on the time elapsed since the champagne was first poured, which determines how much gas is still left dissolved in the liquid, a bubble's journey through champagne to the surface lasts anywhere from 1 to 5 seconds.

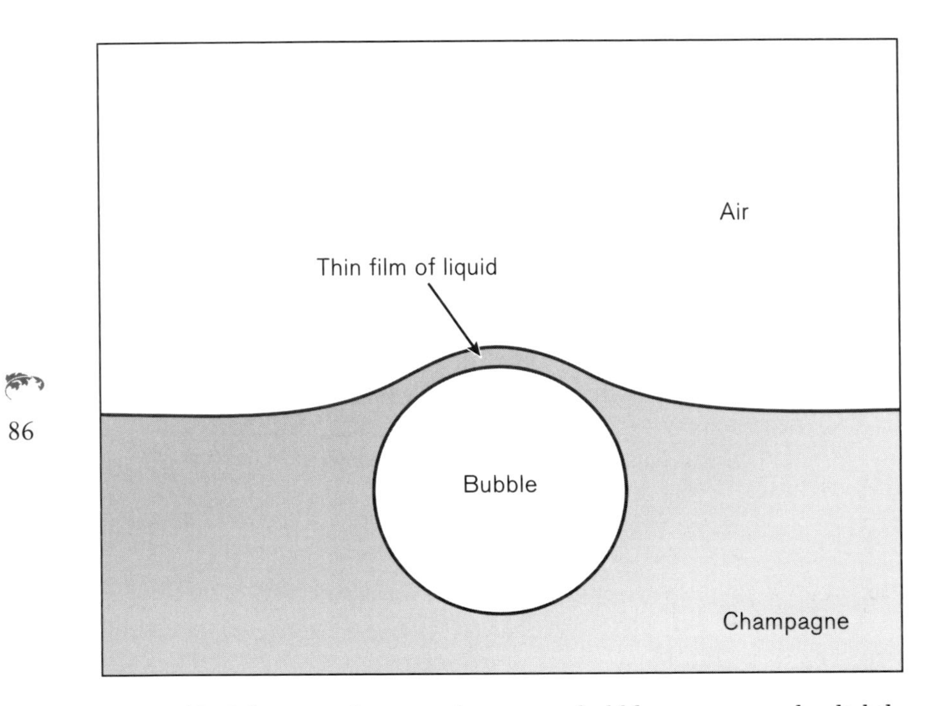

Figure 19 Like an iceberg, a champagne bubble emerges only slightly above the surface of the liquid.

from a balance between buoyancy (which tends to make the bubble emerge from the surface) and a capillary force proportional to the size of the bubble cap (which tends to keep the bubble below the surface). In a flute of champagne, the bubble is so small that buoyancy is completely overpowered by the capillary force. Just like an iceberg in the ocean, a millimetric gas bubble at the free surface of a flute of champagne emerges only slightly above the surface. Most of the bubble volume lies below the liquid surface (Figure 19).

The part of the bubble that pushes above the surface, the *bubble cap,* is essentially a spherically shaped film of liquid that gets thinner and thinner as the liquid drains back into the comparative "ocean" of champagne in the glass beneath it. A bubble cap that has reached a *critical thickness* has reached a thickness below which it will become so thin, fragile, and sensitive to such disturbances as vibrations and temperature changes that it finally ruptures. This thickness is about 100 nanometers, that is, only 0.1 micrometer (or 1/10,000 of a millimeter). You can compare this process in a basic way to blowing up a balloon until the rubber becomes so thin that the balloon finally ruptures.

In 1959, two physicists, Geoffrey Ingram Taylor of the University of Cambridge and Fred E. C. Culick of the California Institute of Technology, independently examined the disintegration,

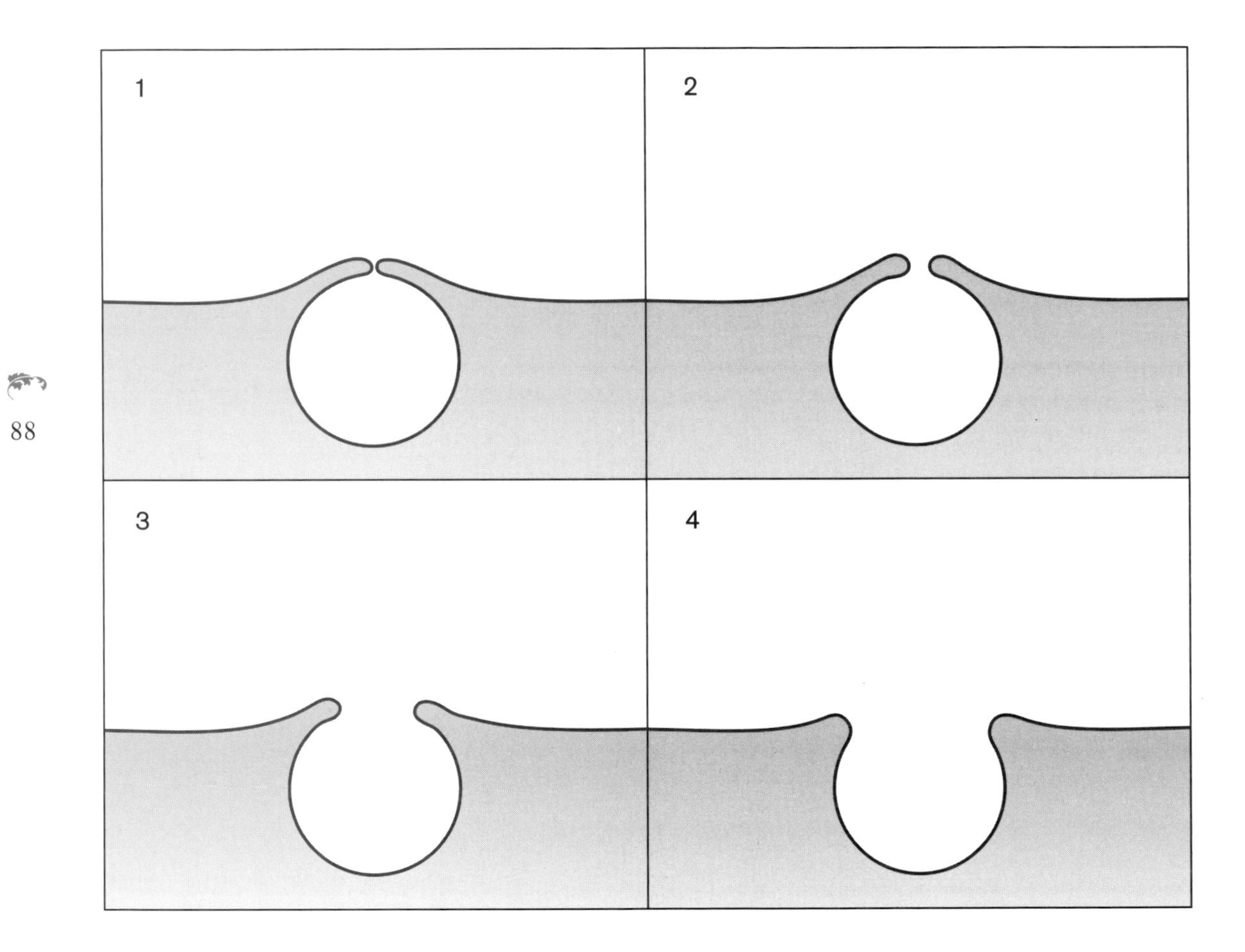

Figure 20 Disintegration of a bubble as it emerges through the surface of a liquid. This process is driven by the forces of surface tension. The hole in the bubble cap propagates at a velocity of about 10 meters per second. This disintegration process is extremely brief. It lasts about only a few tens of microseconds.

or rupturing, process of thin liquid/bubble films. They found that a hole appears in the bubble cap that propagates (i.e., affects the greater area around it) very quickly, propelled by surface tension forces (Figure 20). Typically, the hole in a thin liquid film propagates at a velocity of about 10 meters per second—the speed of the world's best sprinters. As a result, a millimetric champagne bubble's cap disintegrates on a time scale of about 100 microseconds. During this very brief interval, the submerged part of the bubble remains frozen very briefly as a tiny indentation in the liquid surface. Then the submerged part of the bubble collapses as the liquid surface recovers its form and fills in the hollow left by disintegration of the bubble cap.

A reconstructed time sequence illustrating six stages of the collapse of a champagne bubble is shown in Figure 21. Between frames 1 and 2, the thin liquid film, which constitutes the part of the bubble that appears above the surface, has just ruptured. During this extremely brief initial phase, we see the shape of the bubble

1
2
5
6

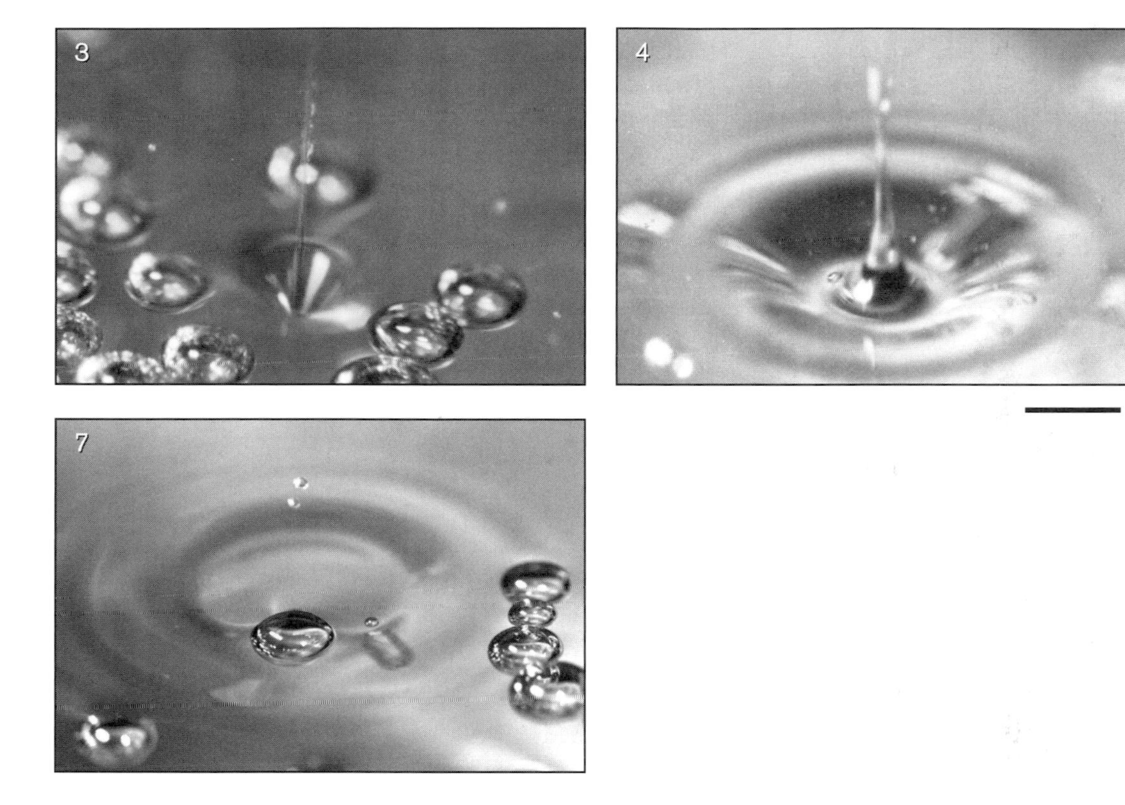

Figure 21 Reconstructed time sequence illustrating the six stages of collapse of a single bubble at the free surface of a glass of champagne. The time interval between each frame is approximately 0.5 millisecond (bar = 1 millimeter). (Photographs by Gérard Liger-Belair.)

and how a nearly millimetric open cavity remains in the liquid surface after the bubble cap ruptures.

While collapsing, the bubble cavity gives rise to a high-speed liquid jet above the free surface of the champagne (frames 3 and 4). The projection of this high-speed liquid jet is caused by pressure gradients around the open cavity formed at the free surface of the champagne when the bubble cap disintegrates. Immediately after rupture of the bubble cap, the region around the cavity becomes an area of positive curvature. This causes a ring of high pressure around the sides of the open cavity. At the same time, the underside of the cavity becomes a region of negative curvature, and this area is a comparatively low-pressure zone. As a result of this disparity, fluid is drawn rapidly from the sides to the bottom of the cavity, where it collides on the axis of symmetry. This collision produces a region of high pressure on the underside of the cavity that serves to push fluid upward in a liquid jet (Figure 22). Incidentally, you

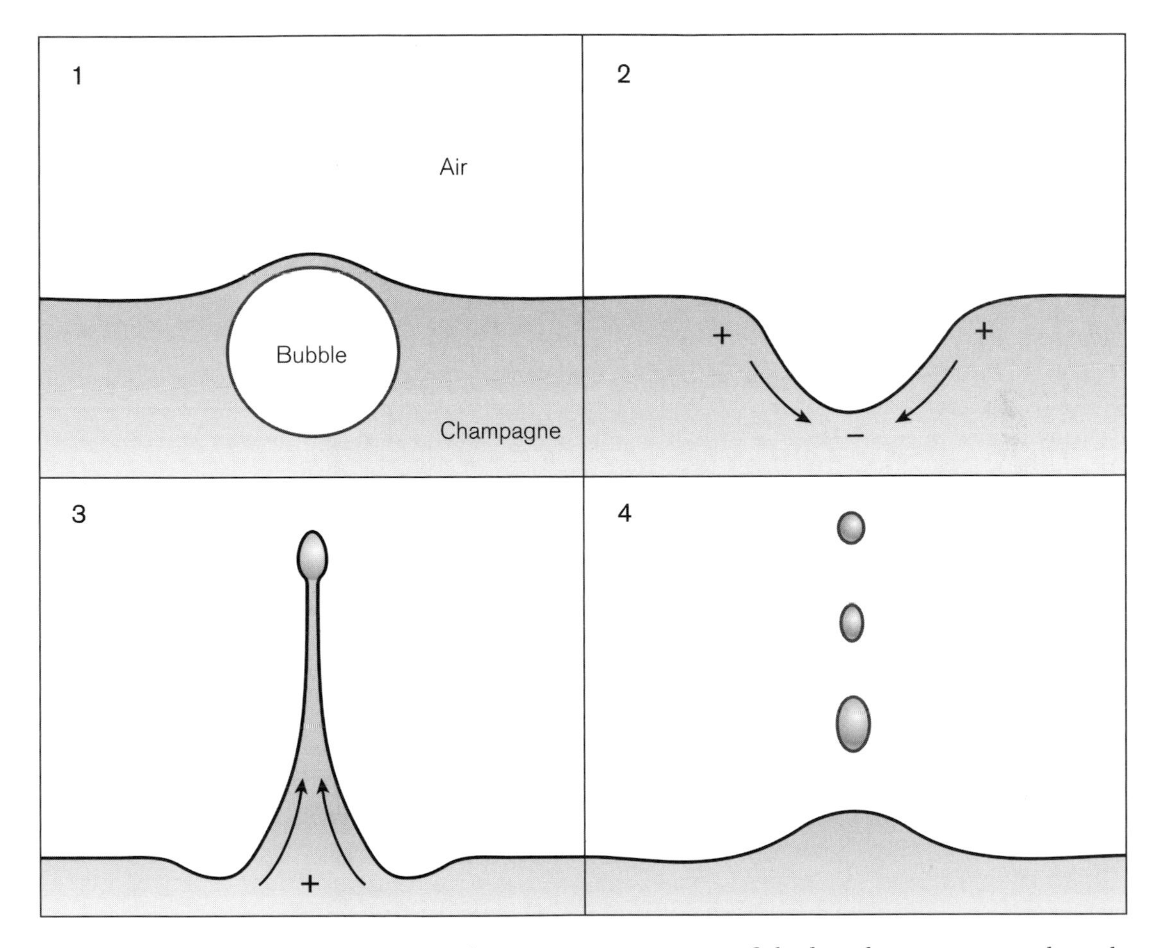

Figure 22 Schematic representation of the liquid streams around a collapsing bubble cavity. These streams lead to the projection of a tiny jet of liquid above the champagne surface. Plus (+) and minus (−) signs indicate pressures above and below atmospheric pressure.

may notice an extremely small bubble (around 100 micrometers) lingering at the base of the liquid jet in frame 4; this bubble is probably a pocket of air trapped in the liquid during the larger bubble's collapse.

The upward-projecting liquid jet created by the bubble's rupture then becomes unstable. A wave known as the *Rayleigh-Plateau instability* (Figure 23) develops along the jet (in frame 5). In frame 6 this instability finally breaks the liquid jet into droplets called *jet drops.* The combined effects of inertia and surface tension give detaching jet drops varied and often amazing shapes. This is exactly the same phenomenon that breaks a thin stream of water trickling from a dripping faucet into droplets. Each millimetric bubble that collapses at the liquid surface gives rise to about five or more jet drops. Finally, in frame 7, droplets ejected by the parent bubble recover a quasi-spherical shape. Due to surface excitations following bubble collapse, capillary wave trains centered on the bursting

Figure 23 Close-up of the upward-projecting jet of liquid as it divides into droplets. A wave known as the *Rayleigh-Plateau instability* propagates along the cylindrical jet of liquid and divides it into droplets called *jet drops* (bar = 1 millimeter). (Photograph by Gérard Liger-Belair.)

bubble propagate at the free surface. On the right side of the central bubble in frame 7, the tiny bubble that was trapped during collapse in frame 4 can still be observed.

A Comparison with Drop Impacts

Harold Edgerton, the twentieth-century master of stop-action photography, invented and developed the electronic flash. He popularized high-speed events by photographing everything from flying bullets and athletes on the move to drops of liquid impacting surfaces. His most famous snapshot, the coronet made by a drop of milk, is familiar to millions of people throughout the world and has become an icon and hallmark of the juncture between pure science and modern art. Even Walt Disney Studios was inspired by Edgerton's photographs; Disney cartoonists applied this image of the drop of milk to the movie *Bambi* in an effort to make raindrops appear more realistic.

Hervé Lemaresquier, a friend and colleague of mine from the University of Reims, reproduced the pioneering work of Edgerton and froze the different stages of a drop of milk impacting the free surface of a container of milk (Figure 24). The liquid jet that follows a bubble collapse strikingly resembles, in miniature, the impact of a droplet on the surface of a pool of liquid. Shape details of the two different liquid jets, one produced during bubble collapse and the other during a drop impact, are displayed in Figure 25. Despite noticeable differences in time and length scales, hydrodynamic structures arising after a drop impact are clearly very similar to those that result from a bubble collapse.

Jet Drops, Sensation, and Flavor Release

Hundreds of bubbles are bursting every second during the first few minutes following the pouring of a flute of champagne; the liquid surface is literally spiked with these conical structures.

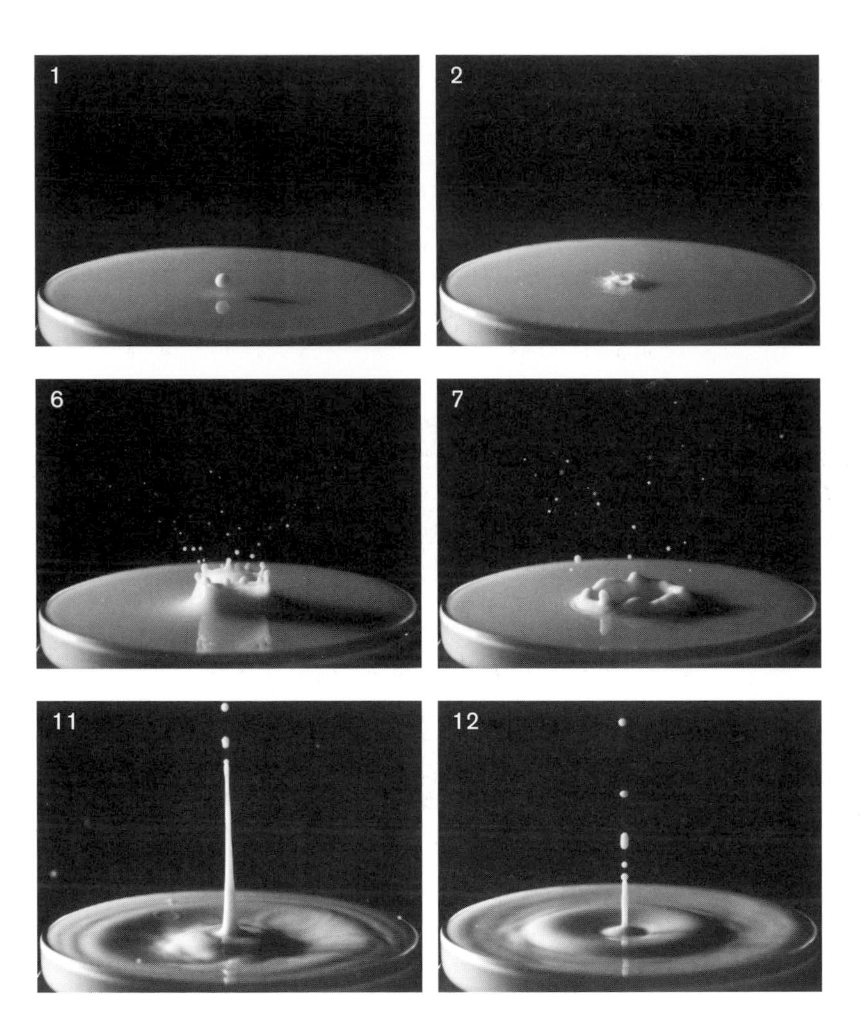
1
2
6
7
11
12

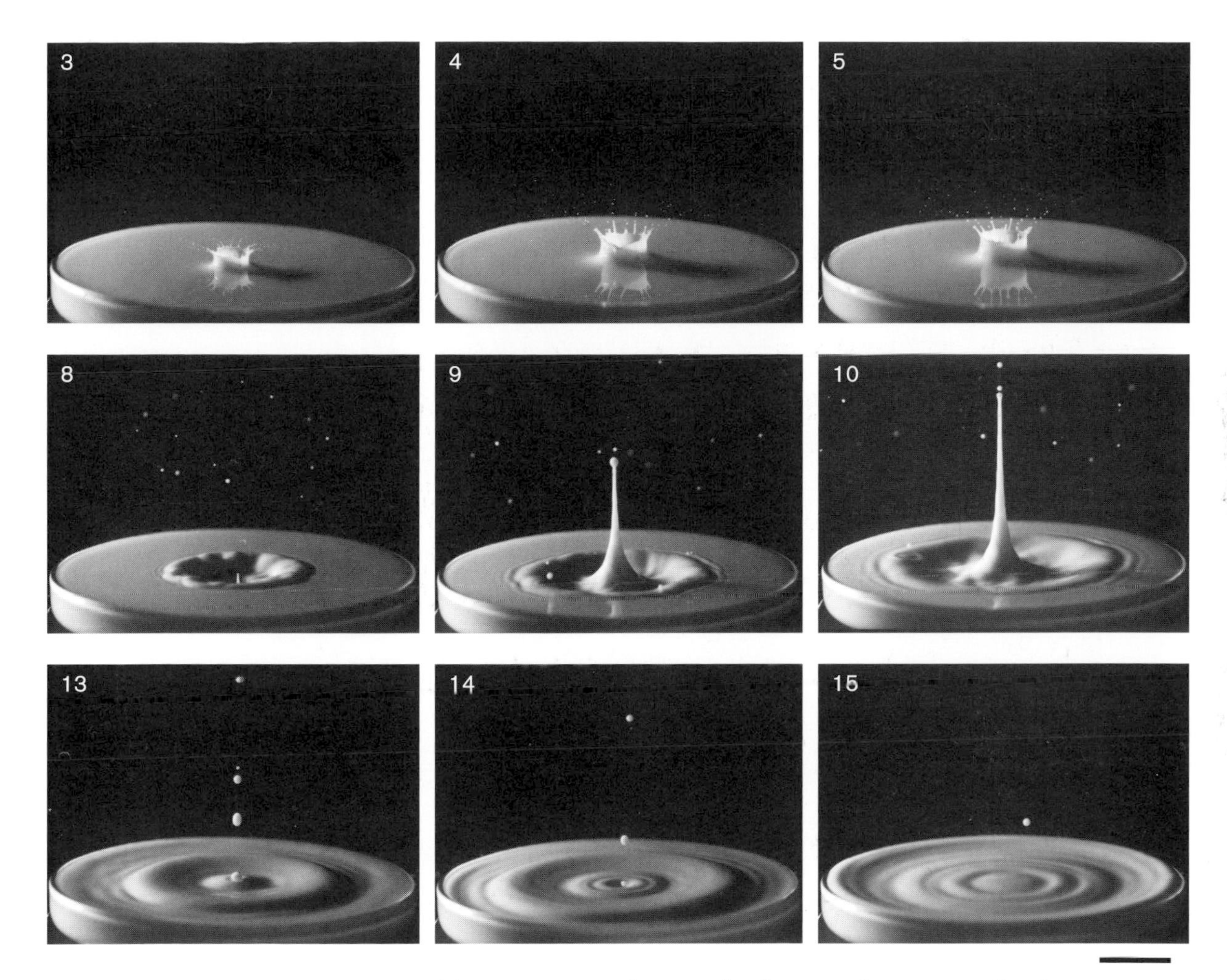

Figure 24 Time-sequence photographs showing the impact of a drop of milk on the surface of a milk-filled container (bar = 1 centimeter). (Photographs by Hervé Lemaresquier.)

Figure 25 Close-up photographs of the two liquid jets that appear around 1 millisecond after a champagne bubble collapses (left) and around 100 milliseconds after a milk drop falls onto the surface of a milk-filled container (right). Despite noticeable differences in time and length scales, the hydrodynamic structures are surprisingly similar and look like tiny Eiffel towers. (Photographs by Gérard Liger-Belair and Hervé Lemaresquier.)

Beyond aesthetic considerations, though, bubbles bursting at the free surface of the liquid impart a sensual feeling to sparkling wines, beers, and many other beverages. The characteristic "mouth feel" of champagne is an important part of drinking it and contributes to the pleasure of tasting. Jet drops are ejected up to several centimeters above the surface with a powerful velocity of several meters per second. In the very first few seconds after pouring, hundreds of bubbles are bursting every second at the liquid surface, and hundreds of tiny jet drops about 100 micrometers in size are launched—in the form of a refreshing spray that pleasantly tickles the taster's face. Indeed, if you wear glasses and take a sip from a freshly poured flute of champagne, you surely have noticed the light spray that forms instantaneously before your eyes. The few centimeters the liquid jets reach into height are just enough to reach our *nociceptors* (the scientific term for very sensitive nerves that act as pain receptors) in the nose; these are thus highly stimulated during champagne tasting, as are receptors in the mouth when bubbles burst over the tongue.

In addition to these mechanical stimulations, bubbles bursting at the surface play a major role in flavor release. Many aromatic compounds in carbonated beverages show surface activity, including alcohols (such as ethanol, butanol, pentanol, and phenyl-2-ethanol), some aldehydes (such as butanal, hexanal, and hexenals), and organic acids (such as propionic and butyric acids). Bubbles rising and expanding in the liquid act like tiny elevators for surface-active and potentially aromatic molecules, carrying them along as they make their way up. Because of this, such molecules gather progressively in much higher concentrations at the surface of champagne than in the bulk of the liquid. Thus, when bubbles collapse at the free surface and cause jet drops, they radiate a cloud of tiny droplets into the air and over the tip of your nose, and these droplets contain a large concentration of potentially aromatic molecules. This fragrant event also creates the delightful and complementary effect of highlighting the flavors of champagne experienced on the tongue and overall causes a heightened sensual experience of the wine for the taster.

Because of these flavor- and scent-carrying bubbles, there is no need to swirl champagne as you would still wines. Rising and bursting bubbles do all the work for you, bringing flavors and aromas above the liquid surface directly to your senses.

A Parallel with the "Fizz" of the Ocean

A nice parallel also can be drawn between the fizz in a glass of bubbly and the "fizz" of the ocean. The major sources of bubbles in the upper ocean are the breaking action of waves and rain impacting on the sea surface, both of which trap air within the seawater. When bubbles created by waves and rain burst at the surface of the sea, they form jet drops just like in champagne. In the early seventies, oceanographers found that droplets ejected into the air at the ocean surface contained much higher numbers of particles and greater amounts of surface-active materials than those found in the bulk of the water directly below; essentially, the jet drops were skimming off the surface layer of the ocean to eject it into the atmosphere, just like champagne jet drops skim off the aromatic particles from the surface of the wine. While bursting bubbles in champagne act to create a sensual effect, bubbles bursting at the sea surface are actually the most significant source of sea salt particles—and these particles act as condensation nuclei to create clouds.

The Fizzing Sound of Champagne

As suggested by the high-speed photographs of bursting bubbles, a lot of energy is released during the collapse of a champagne bubble. Some of this energy is used to create the upward liquid jet; some, however, is also released into the atmosphere in the form of a tiny auditory shock wave. The fizzing sound of champagne is therefore the sum of thousands of individual audible "pops" at the surface of the foam. Nicolas Vandewalle and his colleagues from the University of Liege in Belgium studied the way bubbles collapse in a head of foam made by mixing water and soap. (They had difficulty working directly with champagne because the foamy head of a freshly poured glass of champagne collapses too quickly.) In a special anechoic chamber* the Belgian researchers recorded (with a high quality microphone) the crackling sound made by bubbles collapsing close to each other. They quickly realized that bubbles do not pop independently of one another. The collapse of a bubble strongly affects its surrounding neighbors—which may then collapse in turn, causing a chain reaction throughout the liquid. Scientists call this chain reaction *avalanche behavior*, comparing it to snow

*An anechoic chamber is designed to provide a pure environment that is free of all external noise and vibration.

avalanches in which a falling packet of snow can destabilize other packets of snow and cause a whole mountainside to slide away.

A typical recording of the acoustic activity is displayed in Figure 26. If the bursting events happened at a constant rate while the foam collapses, we would expect to hear a constant and uniform noise similar to *white noise*, like radio static. Instead, champagne seems to *crackle* and is not at all monotonous. The acoustic signal is

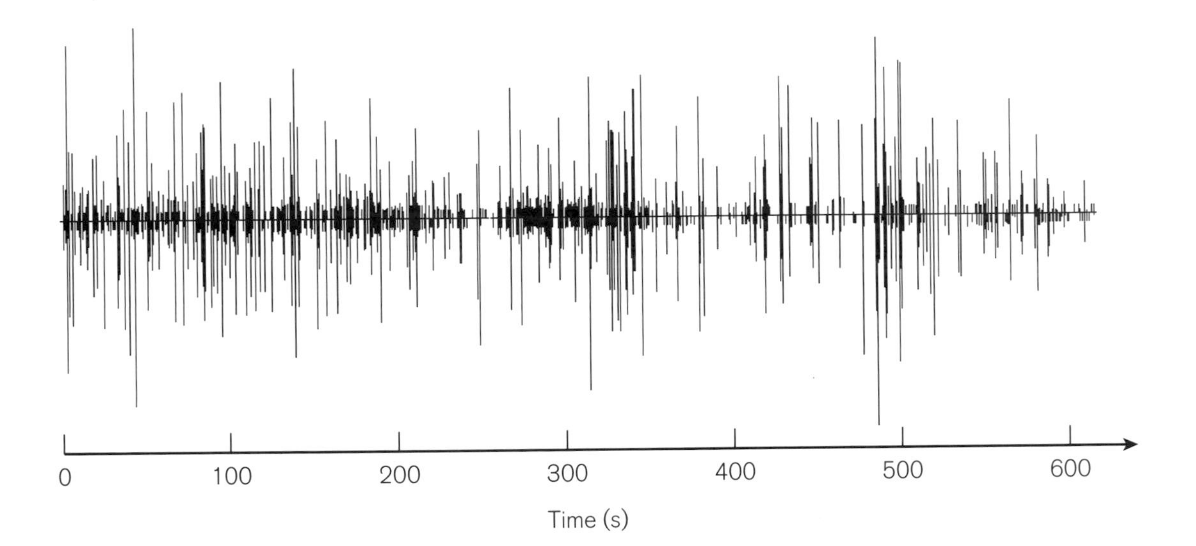

"spiky," and it changes unpredictably in volume as time passes, depending on the amplitude of the chain reactions induced by bursting events. The result? A characteristically capricious fizzing sound that further enriches the sensual experience of champagne, sparkling wines, and beers for the taster.

When Champagne Bubbles Dress Up Like Flowers

Close observation of the bubbles collapsing at the free surface of a glass of champagne also revealed another unexpectedly lovely phenomenon. While snapping pictures of the jet drops caused by champagne bubbles collapsing at the liquid surface, I also accidentally

Figure 26 (Left) Acoustic activity recorded while the foamy head of a mixture of soap and water collapses. Reprinted from Vandewalle et al. (2001) with permission of the authors and of the American Physical Society. © 2001 American Physical Society.

took some pictures of bubbles collapsing close to one another. A few seconds after pouring and after the collapse of the foamy head, the surface of a champagne flute is covered with a layer of bubbles—a sort of bubble raft in which each bubble generally is surrounded by six neighboring bubbles (Figure 27). When the bubble cap of a bubble ruptures and leaves an open cavity at the free surface, adjacent bubble caps are sucked toward this cavity, creating charming flower-shaped structures (Figures 28 through 35). A time sequence of the whole process is displayed in Figure 36. The tiny air bubble entrapped during the collapsing process is seen clearly in frames 5 and 6 of Figure 36.

It is worth noting that despite this violent sucking process, bubbles adjacent to those which burst were never found to rupture and collapse in turn. The dynamics of bubbles collapsing in the champagne bubble raft therefore are rather different from the dynamics of bubbles collapsing in the foamy head that forms when

Figure 27 Oblique view of the bubble raft at the surface of a freshly poured flute of champagne (bar = 1 millimeter). (Photograph by Gérard Liger-Belair.)

FOLLOWING PAGES

Figure 28 (Left) View looking down at the bubble raft at the free surface of a flute of champagne a few seconds after pouring (bar = 1 centimeter). *(Right)* Detail of two flower-shaped structures at the free surface of a flute of champagne a few seconds after pouring. The flower-shaped structures (circled in white in left-hand photo) result from the entrapment of surrounding bubbles in the wake of collapsing bubble cavities (bar = 1 millimeter). (Photographs by Gérard Liger-Belair.)

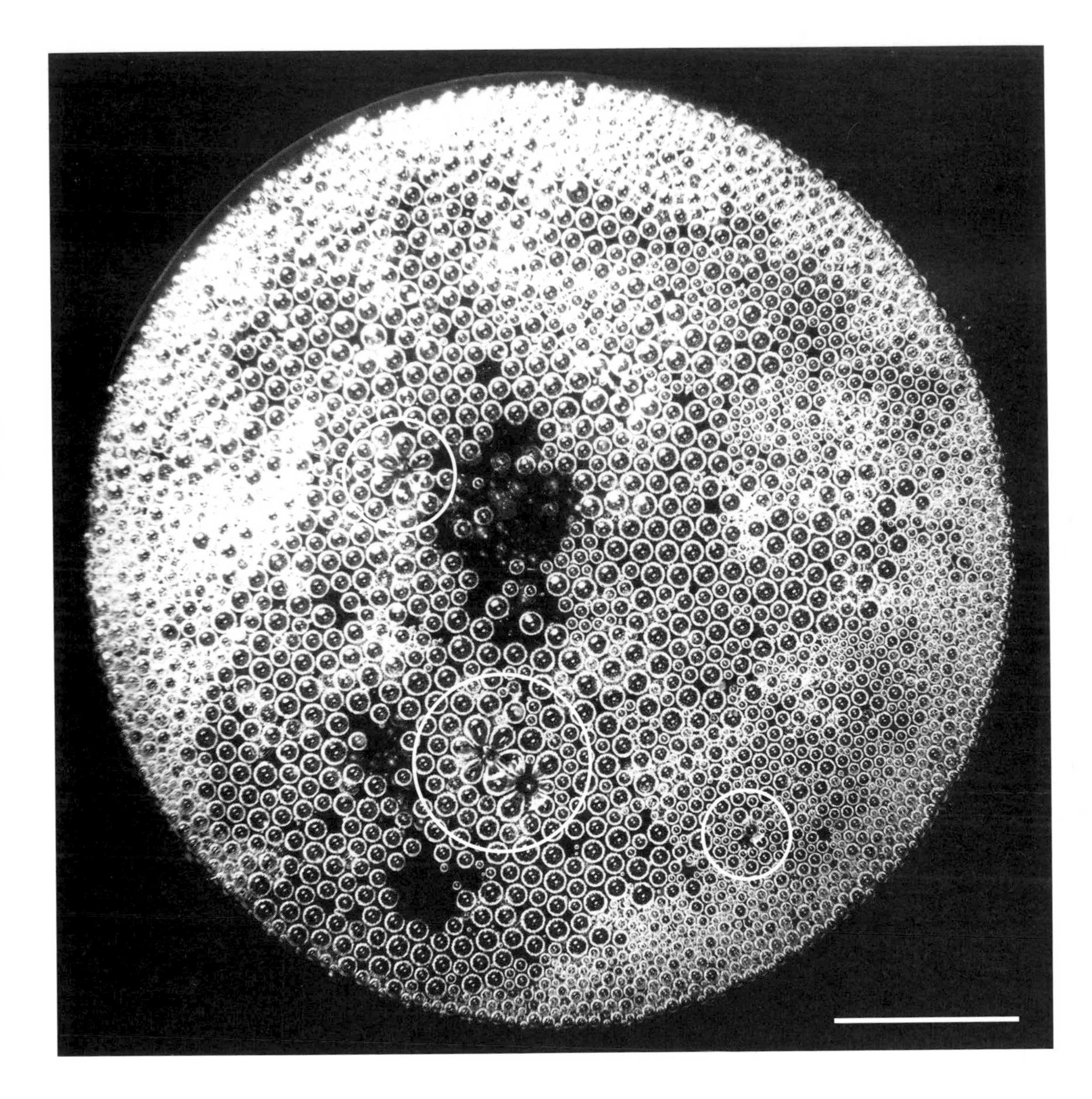

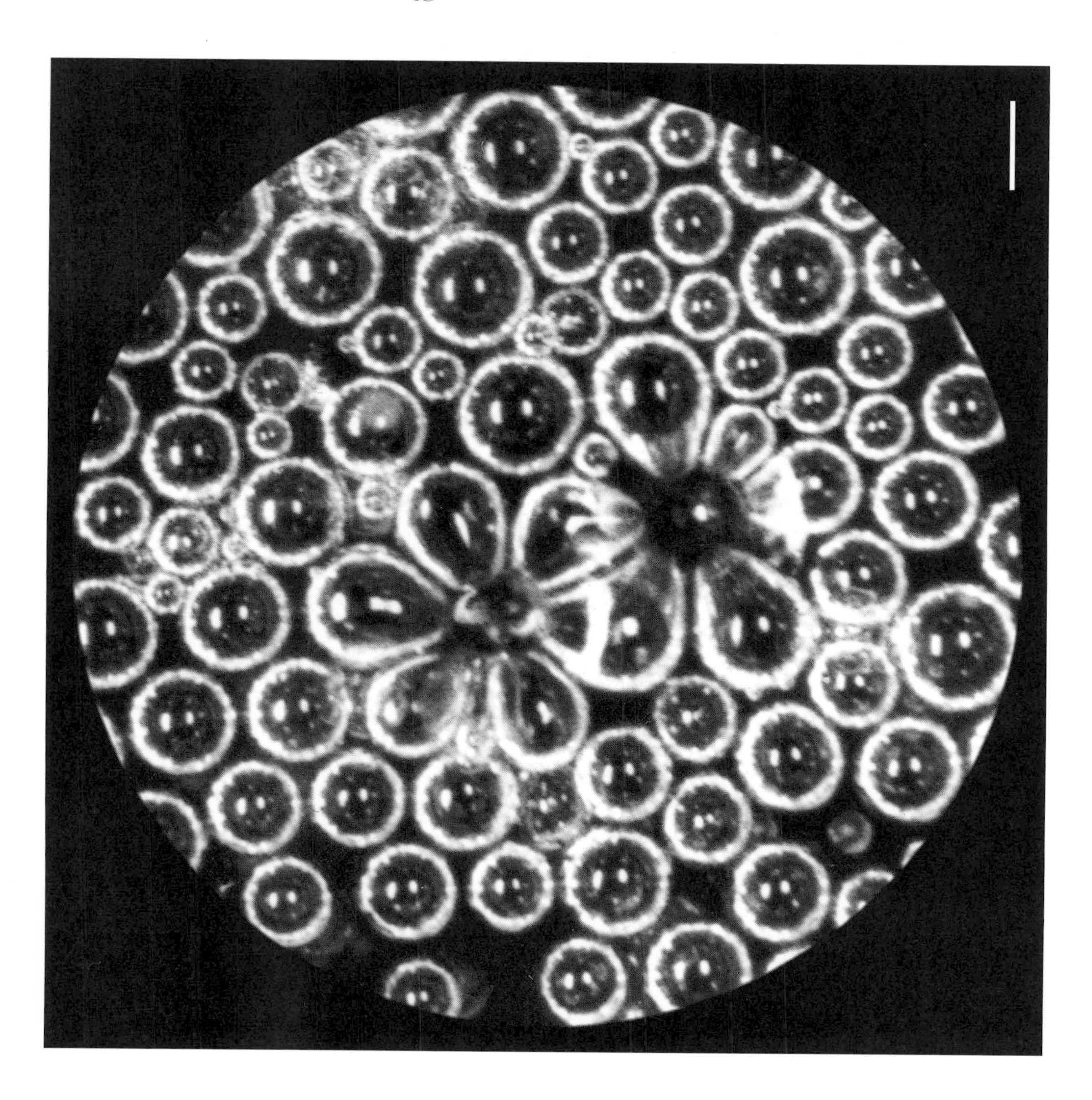

Figure 29 (Top) Oblique view of the collateral effects on adjoining bubble caps of a collapsing bubble (bar = 1 millimeter). (Photograph by Gérard Liger-Belair.)

Figure 30 (Bottom) Close view of the same process as in Figure 29 (bar = 1 millimeter). (Photograph by Gérard Liger-Belair.)

Figure 31 Even closer view of the same process as in Figure 29 (bar = 1 millimeter). (Photograph by Gérard Liger-Belair.)

Figure 32 Extreme close-up view of the same process as in Figure 29. The violence of the sucking process can be seen clearly in this photograph (bar = 1 millimeter). (Photograph by Gérard Liger-Belair.)

Figure 33 Close view of a bursting event in a champagne bubble raft (bar = 1 millimeter). (Photograph by Gérard Liger-Belair.) © 2003 American Chemical Society.

Figure 34 Even closer view of a bursting event in a champagne bubble raft (bar = 1 millimeter). (Photograph by Gérard Liger-Belair.)

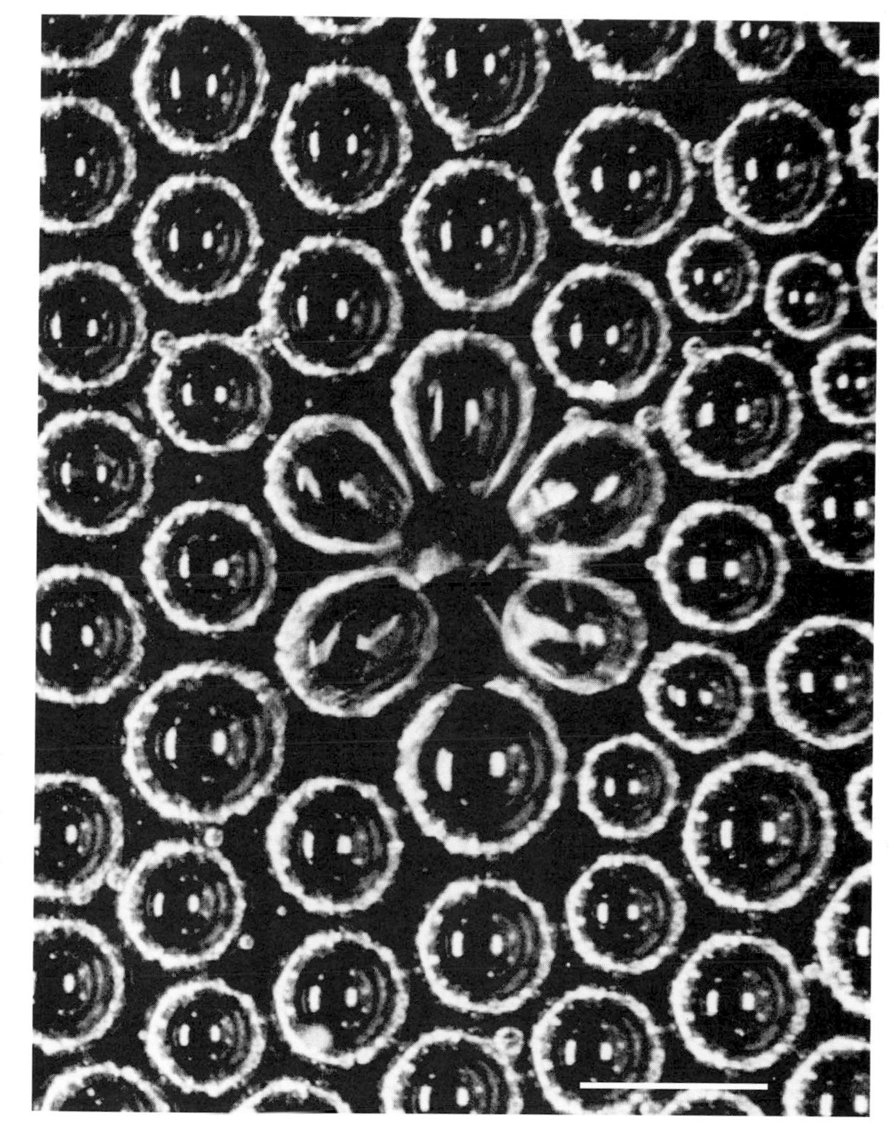

Figure 35 View from above of a bursting event in a champagne bubble raft (bar = 1 millimeter). (Photograph by Gérard Liger-Belair.)

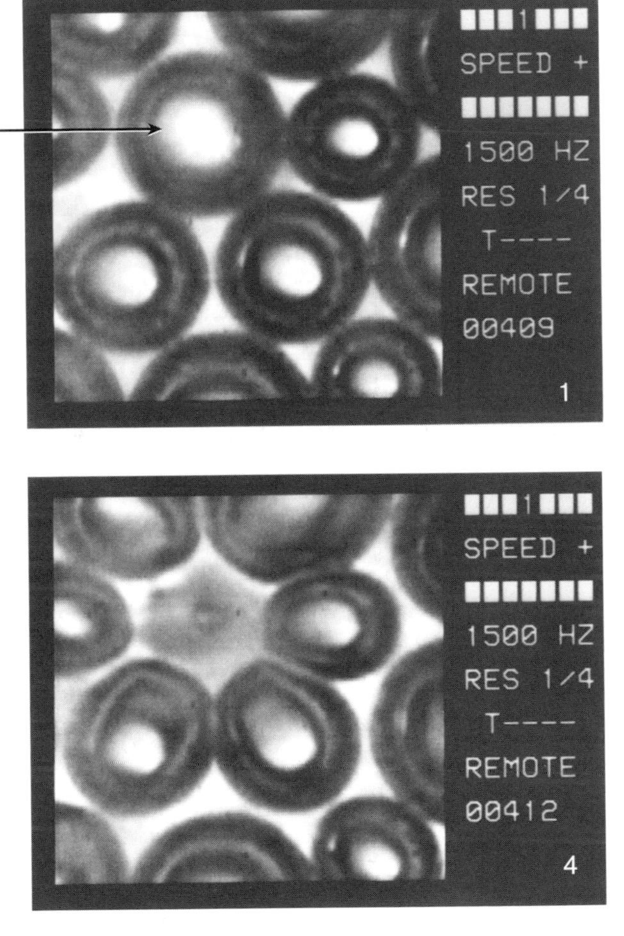

Figure 36 Time-sequence photographs of the collapse of a bubble in the bubble raft of a flute of champagne taken with a high-speed video camera filming at 1500 frames per second. The time interval between each frame therefore is 667 microseconds. Paradoxically, adjoining bubble caps are pulled down rather than pushed up by the forces released by the bursting of the central bubble (bar = 1 millimeter). © 2003 American Chemical Society.

SPEED +
1500 HZ
RES 1/4
T----
REMOTE
00410
2
SPEED +
1500 HZ
RES 1/4
T----
REMOTE
00411
3
SPEED +
1500 HZ
RES 1/4
T----
REMOTE
00413
5
SPEED +
1500 HZ
RES 1/4
T----
REMOTE
00414
6

champagne is poured and where avalanches of bubble bursts are observed. However, bubbles in touch with collapsing ones do experience deformations due to the stress exerted on them by the collapse of the neighboring bubble.

The petals of the flower-shaped structures we see at the surface of a glass of champagne are actually undergoing significant shear stresses* during the sucking process caused by the collapse of the bubble around which they are clustered. During the sudden stretching process caused by shear stresses, the actual surface area of adjacent champagne bubble caps significantly increases, and stresses on adjoining bubbles also increase. When the central bubble finally bursts, adjoining bubble caps absorb the energy released during collapse as tiny "air bags" would do, and they store this energy in the thin liquid film of their emerging bubble caps. This

* The liquid layers inside the deformed bubble caps (or the petals of the flower-shaped structure) are stretched, as suggested by the photographs. The physical quantity that measures the intensity of this stretching is defined as the *shear stress*.

eventually leads to higher stresses around these bubble flowers than you would find around single collapsing bubbles.

Fewer Bursting Bubbles Over Time

Effervescence is generous just after pouring, and collapsing bubbles launch thousands of golden droplets into the air, conveying the champagne's enticing aromas and flavors to the nose. As time passes, though, bursting events become less frequent. Several tens of seconds later, the cloud of tiny jet drops above the liquid surface cannot be observed any more despite the fact that nucleation sites on the glass wall still supply the liquid surface with "fresh" bubbles. Why is this?

Once again, the answer can be found at the molecular level. After pouring, the thousands of rising bubbles bring the surfactants in the champagne to the liquid surface, where they accumulate (Figure 37). As a result, surfactants progressively invade the

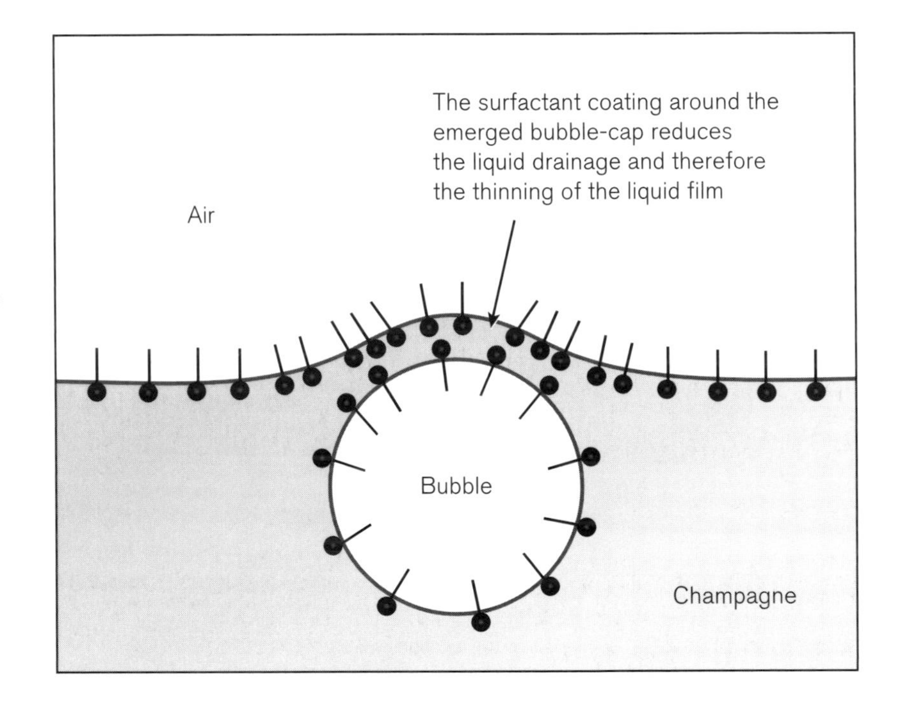

Figure 37 Surfactants progressively decrease the liquid flows inside draining bubble caps and finally prevent the bubbles from bursting.

bubble caps at the liquid surface, coating the bubbles. This surfactant coating on the emerging bubble caps has the same effect as that around ascending bubbles. The surfactant molecules interlock with each other and with the surrounding liquid molecules, strengthening the surface of the bubble and reducing the velocity of the liquid flows in the films of bubble caps. In turn, the liquid inside the thin film of the emerged bubble cap drains more slowly, which extends the lifetime of the bubble (Figure 38). Champagne proteins thus develop a long lasting collar at the periphery of a flute (unlike fizzy waters, for example, where bubbles at the surface are short-lived because of a lack of proteins to rigidify them).

The beneficial role played by the proteins and glycoproteins of champagne nevertheless may be ruined instantaneously by the presence of even minute traces of fatty molecules at the liquid surface. When a drop or molecule of fat touches the liquid membrane of a bubble, it spreads across it very quickly and simultaneously

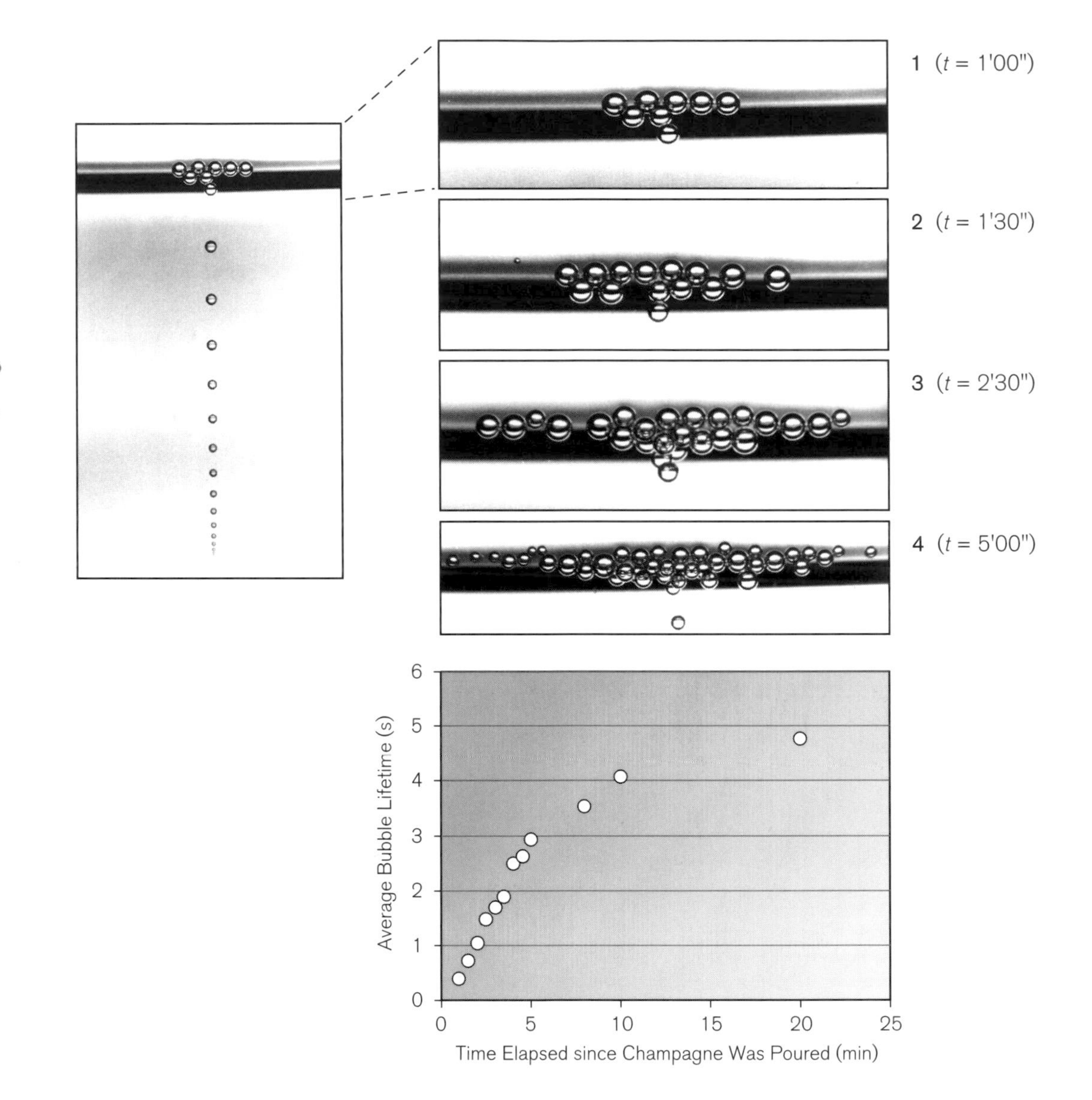
1 (t = 1'00")
2 (t = 1'30")
3 (t = 2'30")
4 (t = 5'00")
Average Bubble Lifetime (s)
0
1
2
3
4
5
6
0
5
10
15
20
25
Time Elapsed since Champagne Was Poured (min)

pulls at some of the liquid molecules making up the membrane. This can cause the bubble membrane to become so thin in places that it ruptures. This is the so-called spreading mechanism that is depicted in Figure 39, and it is why eating chips or peanuts while drinking champagne will make the bubble *collerette* deflate immediately. Also, lipstick is made up in part of fatty molecules, and it also will cause the bubbles to collapse rapidly after the first sip.

However, if the long-lasting bubbles rigidified by surfactants go undisturbed by foods or lipstick, they will offer up yet another aesthetically appealing phenomenon. Gaseous carbon dioxide molecules migrate from the inner bubbles toward the atmosphere, diffusing through the thin liquid film of the emerging bubble caps (Figure 40). Consequently, at the liquid surface, long-lasting bubbles deflate and often disappear before bursting. This bubble deflation

Figure 38 (Left) As time passes, the lifetimes of bubbles increase. (Photographs by Gérard Liger-Belair.) © 2003 EDP Sciences.

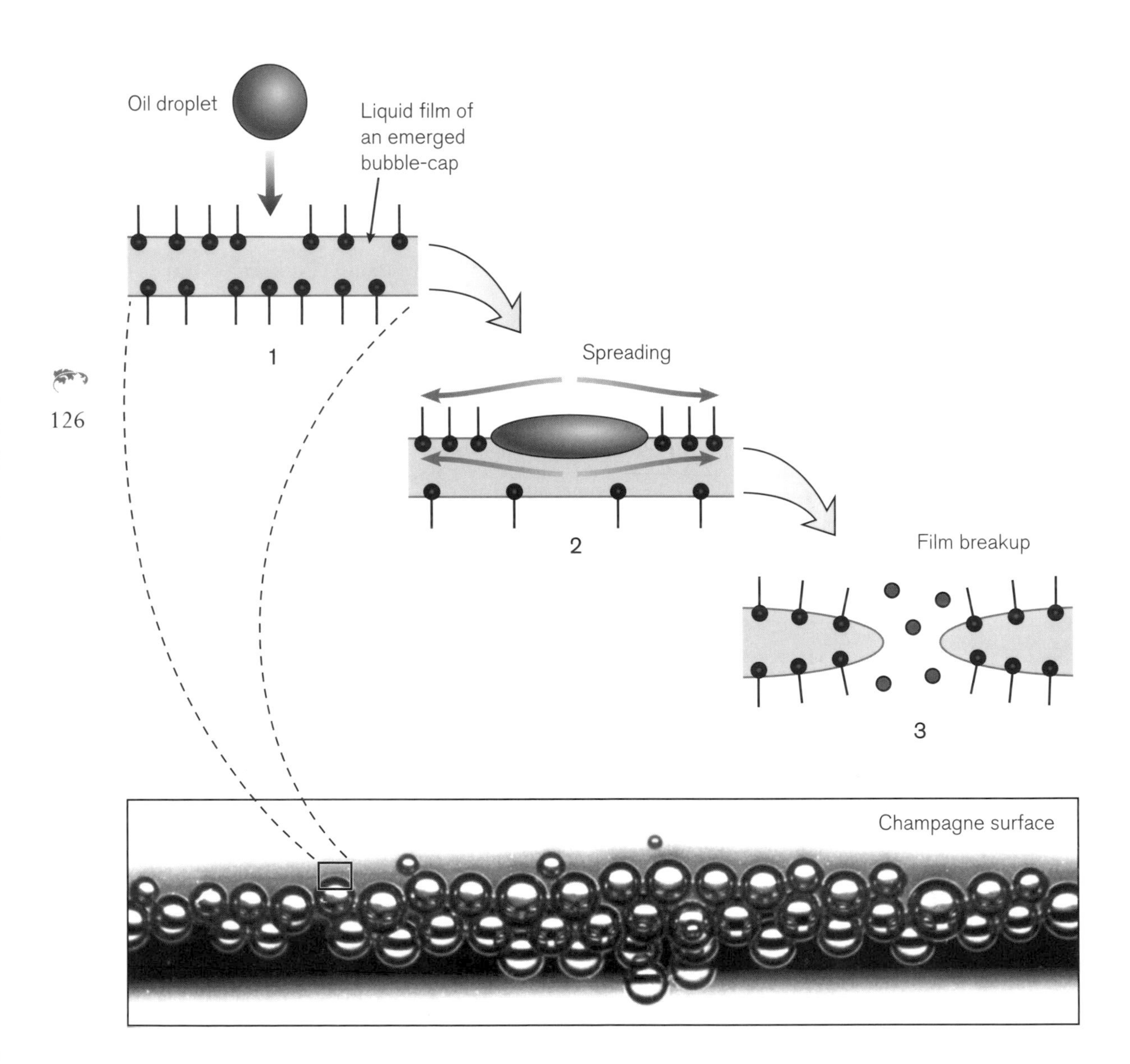
Oil droplet
Liquid film of an emerged bubble-cap
1
Spreading
2
Film breakup
3
Champagne surface

phenomenon is also known as *Ostwald ripening* and may be observed by the naked eye. This is why bursting events become more and more rare. Long after pouring, bubbles at the liquid surface remain and follow the liquid motion. And if you do not disturb the play of bubbles within the liquid by swirling the flute, you will see that they organize themselves into a slow two-dimensional vortex that looks delightfully like a galaxy (Figure 41). These "bubble stars" at the champagne surface rotate in such a pattern simply because of the presence of the circular wall of glass containing them, which obliges the bubbles to confine their spirals to its boundaries.

And now, at last, if you haven't yet, toast yourself and take a long sip from your glass. While you do, keep in mind that you are

Figure 39 (Left) Schematic representation of the spreading mechanism caused by grease that accelerates bubble rupture at the surface of a flute of champagne and therefore reduces the average lifetimes of bubbles. (Photograph by Gérard Liger-Belair.)

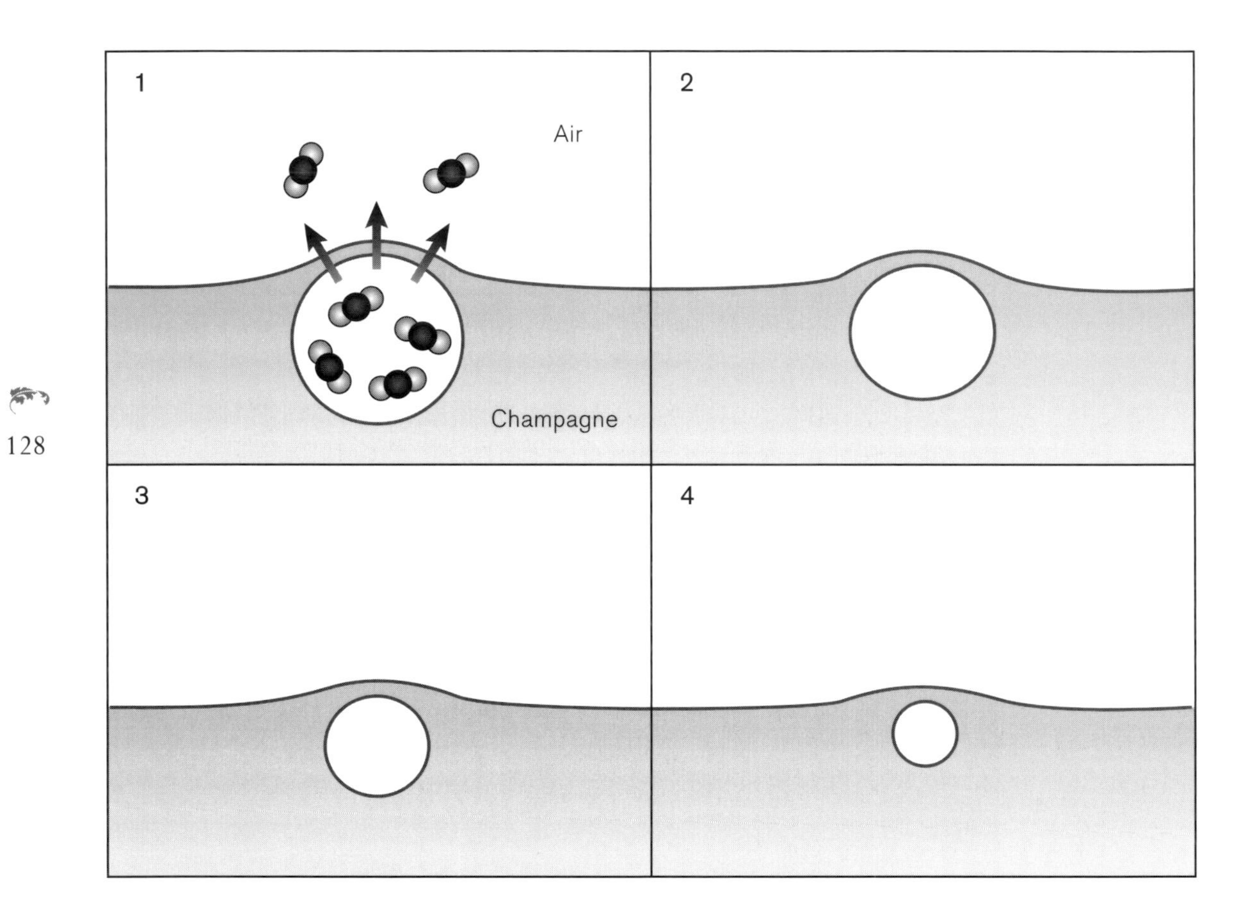

Figure 40 At the liquid surface, long-lasting bubbles rigidified by surfactants shrink progressively because carbon dioxide molecules diffuse from inside the bubbles to the atmosphere through the bubble caps.

now able to see the unseen. As the wine flows and crackles over your tongue, as your nose and mouth respond to the scents, flavors, and sensations so peculiar and unique to the drink, remember that your experience is richer now for the intimate knowledge you have gained about the drink in your hand. Take another sip. How does the surface of the wine feel against your lips now that you know they are scattering the petals of a virtual garden of bubble flowers every time you take a taste? How do the tip of your nose and eyelashes feel now that you know that they have just been assailed by innumerable drops of perfume reaching up from the surface of the glass in bursts and towers of scent? How does the wine appeal to your eyes now that you know the origins of the pearly trains of bubbles rising up through the flute?

The drink in your mouth and in your glass is a symbol fraught with contradictions, one that summons up notions of both decadence and civility, whose creation was a matter of both accident

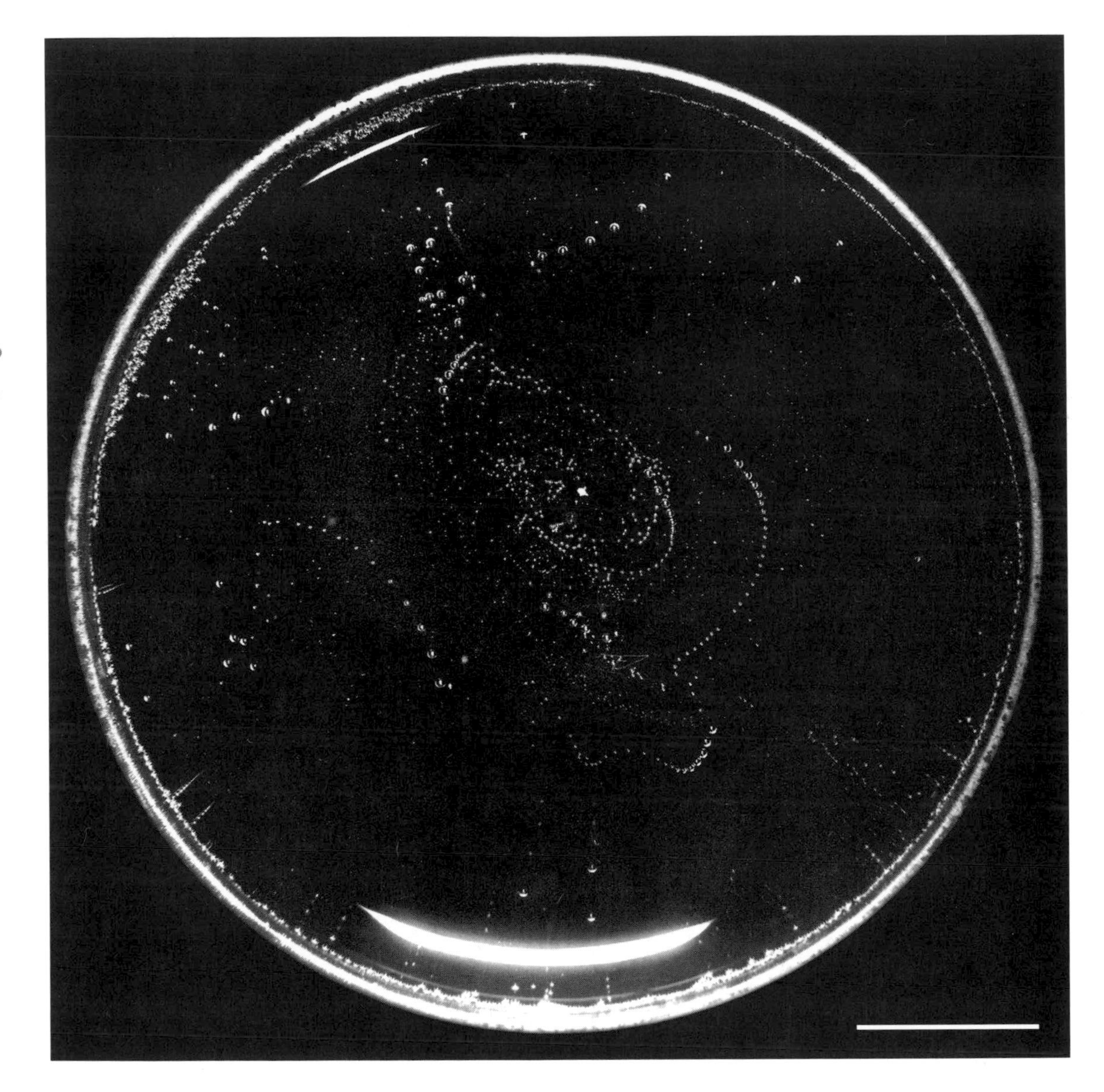

Figure 41 (Left) Long after pouring, bubbles at the liquid surface last a very long time. They organize themselves in slow two-dimensional vortices that, surprisingly, look like galaxies (bar = 1 centimeter). *(Right)* Close-up of a two-dimensional vortex (bar = 1 millimeter). (Photographs by Gérard Liger-Belair.)

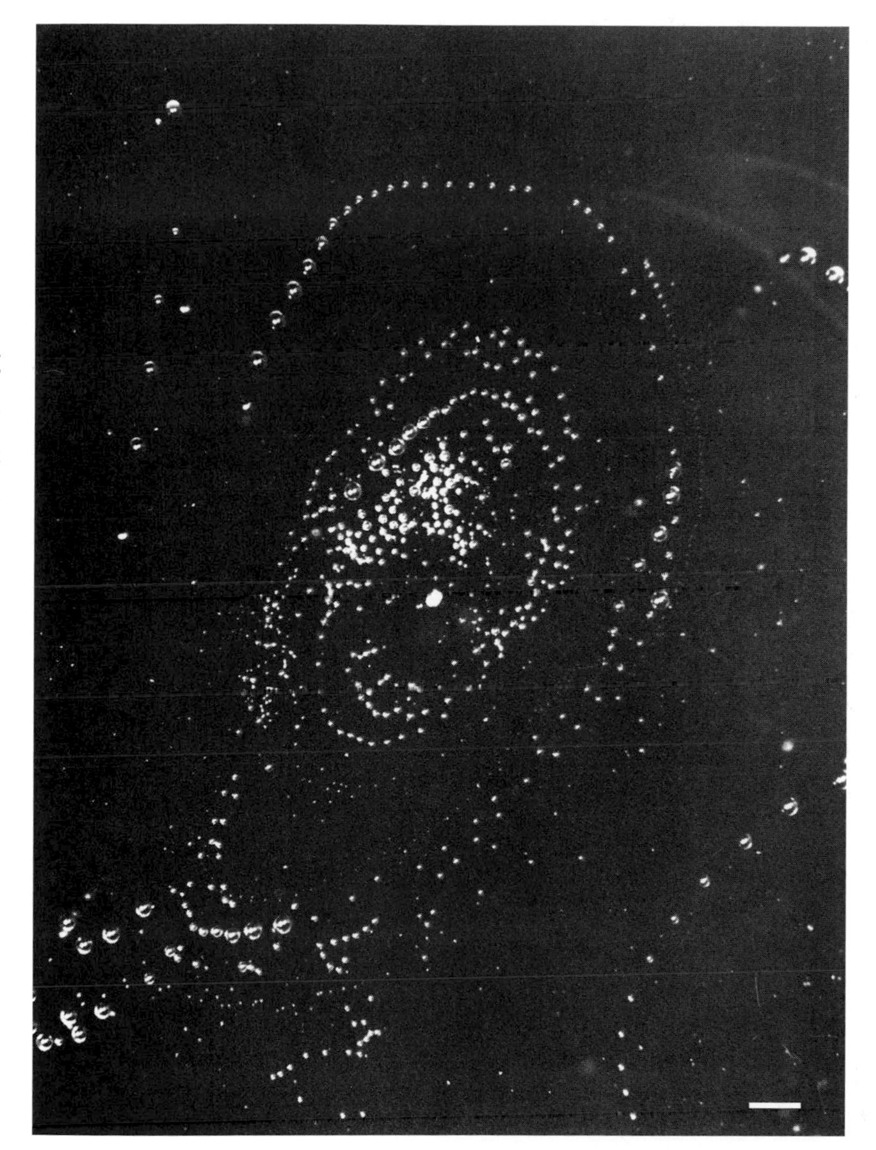

and cultivation, and whose allure can be sensed on countless and varying levels. By reading this book you have refined your sense of taste and broadened your vocabulary for the appreciation of beauty. Your capacity for enjoyment is, I hope, made richer for the time you took to gain insight into the physics and chemistry of champagne effervescence. Your effort does honor not only to the unique drink in your hand but also to yourself. I hope that the experience inspires you to savor all your life's moments just as you would relish every sip of champagne. *Salut!*

AFTERWORD: THE FUTURE OF CHAMPAGNE WINES

Most environmentalists and climatologists now believe that our planet is getting warmer and warmer. Global warming is due to an increase in the magnitude of the greenhouse effect, as well as an increase in solar intensity (due to ozone depletion). The earth experiences a natural greenhouse effect due to trace amounts of certain naturally occurring gases (e.g., carbon dioxide, water vapor, methane, and others). Carbon dioxide is one of the primary greenhouse gases in the atmosphere, and it actually traps outgoing heat and warms the Earth. The *enhanced* greenhouse effect refers to the augmentation of these gases by human activities—for example, the burning of fossil fuels such as oil, coal, and natural gases—that release carbon dioxide into the atmosphere. Scientists can measure the amount of carbon dioxide in the Earth's atmosphere and have discovered that carbon dioxide levels are increasing. In 1958, a U.S. professor of oceanography named Charles Dave Keeling began a systematic study of atmospheric carbon dioxide levels in a remote lava field near Hawaii's Mauna Loa volcano. His results, summarized in a graph known as a *Keeling curve* (Figure 42), clearly show the steady rise in carbon dioxide in the atmosphere—from about 316 part per million by volume in 1958 to about 369 part per million by volume in 2000. And there is every reason to believe that carbon dioxide levels will continue to rise if political and technological solutions are not put in place to inhibit the continued release of greenhouse gases into the atmosphere. Unfortunately, since President G. W. Bush refused to agree to the terms of the Kyoto Protocol in 2002—a United Nations effort to reduce the

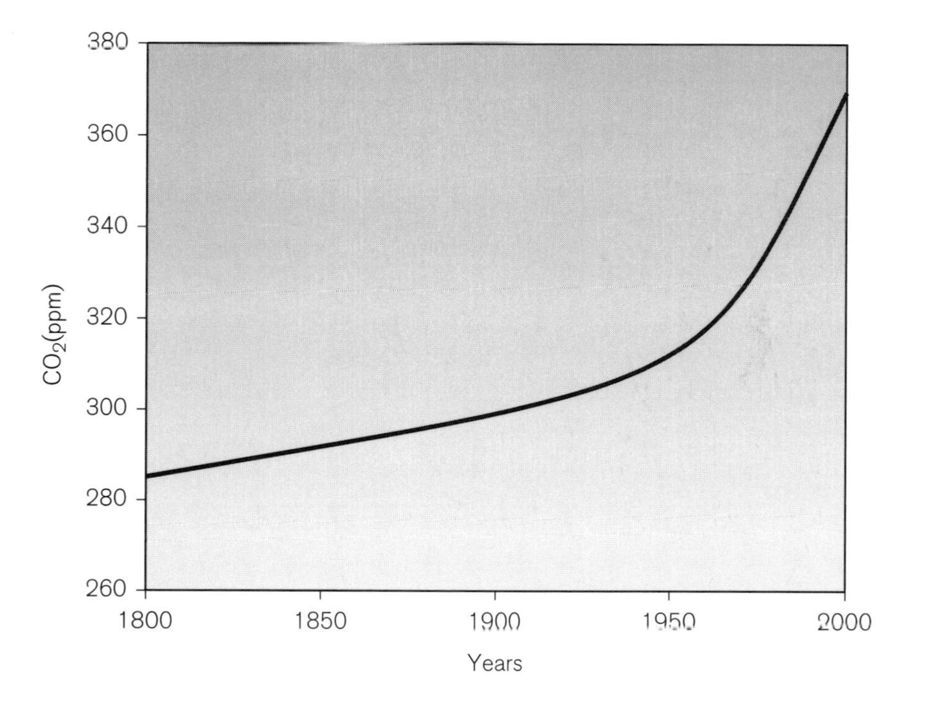

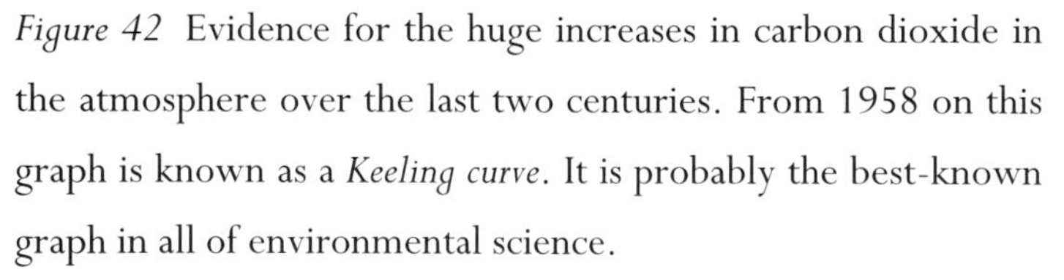

Figure 42 Evidence for the huge increases in carbon dioxide in the atmosphere over the last two centuries. From 1958 on this graph is known as a *Keeling curve*. It is probably the best-known graph in all of environmental science.

amount of greenhouse gases emitted by developed countries—the outlook of many environmentally minded countries on this issue presently remains rather gloomy.

The enhanced greenhouse effect induced by this rise of carbon dioxide into the atmosphere should increase the average temperature on Earth by between 2 and 4° C during the next one hundred years depending on the latitude and the optimism of the different climatic models. Chaotic weather patterns, catastrophic droughts, a significant rise in sea levels, and unusual precipitation are all linked to this increase in the global average temperature. According to some scientists, global warming already has had dramatic consequences on the climatic equilibrium of Earth during the last decade. In December 1999, for example, Europe was swept by violent windstorms. Dozens of people were killed. Millions of trees were blown down. About 10,000 trees in the woodlands at the Versailles Palace, many of them historic, were lost. Scientists said that during the twenty-first century, hurricanes, typhoons, and cyclones in the Indian Ocean and South Pacific are likely to produce higher winds and heavier rains in some areas. Recent computer modeling even suggests that the maximum intensity of cyclones or hurricanes may increase by up to about 10 percent in terms of wind velocity as greenhouse gases increase.

Global warming also undoubtedly will have significant consequences for the grapevine culture in general and for champagne making in particular. Champenois winegrowers strongly believe that the northerly climate and the thick, chalky, limestone-rich subsoil of the Champagne region give champagne its unique style and personality. A cool climate produces light, tart wines that are quite poor as still wines but perfect for top-quality sparkling ones. Chalky soils give the grapes an acidity that allows the full strength of the aroma to develop over longer aging periods. Warmer and longer growing seasons in the Champagne region could seriously modify both the sugar and acid levels of the grapes, and these changes, in turn, could affect the general sensual properties of the wine and its aging potential.

While I was writing this book, during summer of 2003, Europe was in the grip of the worst heat wave since reliable record keeping began about 130 years ago. Consequently, the 2003 harvest came after an extraordinarily hot summer, with temperatures regularly soaring above 35° C. This year, we experienced 40° C for a period of two weeks. The lack of rain and the heat not only caused many grapes on the vines to dry up prematurely, but the hot sun and dry conditions also brought high levels of natural sugar and great concentration to the grapes. The acidity levels of the grapes were high at first, but they decreased as the sugar levels increased—a development that concerned winemakers because they know that a well-balanced champagne needs acidity for good structure and aging. Therefore, in order to preserve the remaining acidity, the grapes were harvested as soon as possible (at the end of August, or three weeks before the usual date). The final result was that the overall production this year was about 50 percent less than in a normal year.

The quality of this year's vintage is difficult to predict because there was nothing normal about this year. The Champagne region is an area that is naturally cooler during the summer months than many of the other wine-producing regions of France and the rest of Europe, so comparatively hot summers usually produce a good, albeit smaller, vintage of still wines for Champagne. However, the unusually severe heat wave included the area of Champagne this summer and may have influenced the region's 2003 vintage in uncertain ways. The heat wave also causes us to wonder whether this exceptional year is a symptom of natural climate variability or the beginning of a large-scale climate change—perhaps a change on an even greater scale than that which caused the first natural fizzing in champagne so long ago. There are many uncertainties that make answering this question difficult, but most climatologists now believe that we are facing the first large-scale climate change induced by human activities and that a lack of constraint on human emissions may change our climate significantly from that which we know today.

If we are at the beginning of a large-scale climate change or global warming, the flavor and taste of Champagne wines could change and possibly suffer; however, thankfully, the consequences on the bubbling behavior should not be severe. Bubbling is a physical process that depends more heavily on exogenous factors, such as the presence of nucleation sites and the cleanliness of the flute rather than on the wine itself. However, the trend toward increasing humidity and rising air temperature suggests that the risk of fungal disease in the vineyards may increase, which eventually could modify the protein compounds of the base wine significantly, and unfortunately, such a change could succeed in modifying the bubbling behavior of the final product. Since proteins act as surfactants that stick to the surfaces of rising bubbles and create rigid coats around them, this change and its effect could modify the lifetime of bubbles at the liquid surface significantly. It was found recently, for example, that grapes infected with *Botrytis cinerea* fungus* produced a champagne with significantly less foaming or bubbling properties than grapes that had not been infected.

Interestingly, some scientists at Oxford University believe that if the climate is beginning to change significantly as a result of human activity and emissions, the wine-friendly climate of the reputed French Bordeaux region will be found in about 50 years in Great Britain, where there are now only struggling vineyards. Perhaps this is just wishful thinking on some level, but in actuality, some regions in the south of England possess the same chalky soil (this limestone-rich soil is the source of the famous white color of

*The fungus *Botrytis cinerea* is a mold produced in very particular conditions of temperature and humidity that pierces grape skins, causing dehydration.

the cliffs of Dover) as that found in the Champagne region. Thus it is not completely unrealistic to think that we eventually may witness the emergence of fine sparkling wines in Great Britain. However, at the present time, we cannot be certain about the precise magnitude of the impact that global warming may have on the future of champagne, its characteristic bubbling, or the wine-producing capabilities of its famed region of origin. We can only hope that the Champagne region will always be known for its namesake wine and that *le vin des rois et le roi des vins* only improves, as it has always done, with age and time.

GLOSSARY

Adsorption The process of trapping molecules at the boundary (frontier) between two phases.

Atmosphere Unit of pressure. Based on the original Torricelli barometer design, one atmosphere (atm) of pressure will force a column of mercury in a mercury barometer to a height of 760 millimeters. Under the cork of a champagne bottle, the pressure of carbon dioxide may reach up to 6 atmospheres. 1 atm = 760 mmHg = 101.3 kPa = 14.7 $lb/in^2 \approx 10^5 N/m^2$.

Buoyancy Also known as *Archimedes' principle*. A body in a fluid, whether floating or submerged, is buoyed upward by a force equal to the weight of the fluid displaced by the body.

Critical radius Minimum radius required (thermodynamically speaking) for a bubble embryo trapped inside an immerged particle to enable dissolved carbon dioxide molecules to migrate from the champagne to the gas pocket, thus beginning the bubble production process. Below this critical radius, bubble formation becomes impossible.

Lumen Hollow cavity inside a structure composed of cellulose fiber that is not completely wetted when champagne is poured and where a trapped gas pocket initiates the repetitive bubble production process. Scientifically speaking, the fiber's lumen acts as a *nucleation site*.

Nucleation The tendency of compounds to cluster in settlements of increasing size. Scientifically speaking, champagne bubbles appear by nonclassic heterogeneous nucleation, that is, from preexisting gas cavities. We talk about classic homogeneous nucleation when bubbles appear directly in the bulk of the liquid.

Rayleigh-Plateau instability Wave instability that develops along a cylindrical trickle of liquid and finally breaks it into droplets. The name is taken from the two famous physicists who studied such phenomena, Englishman Lord Rayleigh and Belgian Joseph Ferdinand Plateau.

Surfactant A commonly used shortened form of "surface-active agent." Surfactants are compounds that have a water-soluble and a water-insoluble part. Consequently, they adsorb at the boundaries between gases and liquids. They are used most often in detergents and soaps as foaming agents.

BIBLIOGRAPHY

Aybers, N. M., and Tapuccu, A. 1969. "The motion of gas bubbles rising through stagnant liquid." *Wärme und Stoffübertragung* 2:118–128.

Ball, P. 2000. "Bottoms up." *Nature Science Update*, March; available at http://www.nature.com/nsu/000302/000302-8.html.

Dussaud, A., Robillard, B., Carles, B., Duteurtre, B., Vignes-Adler, M. 1994. "Exogenous lipids and ethanol influences on the foam behavior of sparkling base wines." *Journal of Food Science* 59:148–167.

Edgerton, H., and Killian, J. R. 1939. *Flash! Seeing the Unseen by Ultra High-Speed Photography*. Boston: Hale.

Harper, J. 1970. "On bubbles rising in line at large Reynolds numbers." *Journal of Fluid Mechanics* 41:751–758.

Harper, J. 1997. "Bubbles rising in line: Why is the first approximation so bad?" *Journal of Fluid Mechanics* 351:289–300.

Jones, S. F., Evans, G. M., and Galvin, K. P. 1999. "Bubble nucleation from gas cavities: A review." *Advances in Colloid and Interface Science* 80:27–50.

Knutson, T. R., Tuleya, R. E., and Kurihara, Y. 1998. "Simulated increase of hurricane intensities in a CO_2 warmed climate." *Science* 279:1018–1021.

Liger-Belair, G. 2003. "The science of bubbly." *Scientific American* 288:80–85; available at http://www.sciam.com/issue.cfm?issuedate=Jan-03.

Liger-Belair, G., and Jeandet, P. 2003. "More on the surface state of expanding champagne bubbles rising at intermediate Reynolds and high Peclet numbers." *Langmuir* 19:801–808.

Liger-Belair, G., and Jeandet, P. 2003. "Capillary-driven flower-shaped structures around bubbles collapsing in a bubble raft at the surface of a liquid of low viscosity." *Langmuir* 19:5771–5779.

Liger-Belair, G., Lemaresquier, H., Robillard, B., Duteurtre, B., and Jeandet, P. 2001. "The secrets of fizz in Champagne wines: A phenomenological study." *American Journal of Enology and Viticulture* 52:88–92.

Liger-Belair, G., Marchal, R., and Jeandet, P. 2002. "Close-up on bubble nucleation in a glass of champagne." *American Journal of Enology and Viticulture* 53:151–153.

Liger-Belair, G., Marchal, R., Robillard, B., Dambrouck, T., Maujean, A., Vignes-Adler, M., and Jeandet, P. 2000. "On the velocity of expanding spherical gas bubbles rising in line in supersaturated hydroalcoholic solutions: Application to bubble trains in carbonated beverages." *Langmuir* 16:1889–1895.

Liger-Belair, G., Robillard, B., Vignes-Adler, M., and Jeandet, P. 2001. "Flower-shaped structures around bubbles collapsing in a bubble monolayer." *Comptes Rendus de l'Académie des Sciences (Paris),* Series 4, 2:775–780.

Liger-Belair, G., Vignes-Adler, M., Voisin, C., Robillard, B., and Jeandet, P. 2002. "Kinetics of gas discharging in a glass of champagne: The role of nucleation sites." *Langmuir* 18:1294–1301.

Lohse, D. 2003. "Bubble puzzles." *Physics Today* 56(2):36–41; available at http://www.physicstoday.org/vol-56/iss-2/p36.html.

Marchal, R., Tabary, I., Valade, M., Moncomble, D., Viaux, L., Robillard, B., and Jeandet, P. 2001. "Effects of *Botrytris cinera* infection on champagne wine foaming properties." *Journal of the Science of Food and Agriculture* 81:1371–1378.

Mougin, G., and Magnaudet, J. 2002. "Path instability of a rising bubble." *Physical Review Letters* 88:014502.

Perkowitz, S. 2000. *Universal Foam: From Cappuccino to the Cosmos.* New York: Walker Publishing.

Shafer, N., and Zare, R. 1991. "Through a beer glass darkly." *Physics Today* 44:48–52.

Stevenson, T. 1998. *Christie's World Encyclopedia of Champagne and Sparkling Wine.* Bath, England: Absolute Press.

Topgaard, D. 2003. "Nuclear magnetic resonance studies of water self-diffusion in porous systems." Ph.D. thesis, Lund University, Lund, Sweden.

Vandewalle, N., Lentz, J. F., Dorbolo, S., and Brisbois, F. 2001. "Avalanches of popping bubbles in collapsing foams." *Physical Review Letters* 86:179–182.

Weart, S. R. 1997. "The discovery of the risk of global warming." *Physics Today* 50: 34–40.

Weiss, P. 2000. "Toasting a burst of discovery about bubbles in champagne and beer." *ScienceNews* 157(19):300–302; available at http://www.sciencenews.org.

Woodcock, A. H., Kientzler, C. F., Arons, A. B., and Blanchard, D. C. 1953. "Giant condensation nuclei from bursting bubbles." *Nature* 172:1144–1145.

Wu, M., and Gharib, M. 2002. "Experimental studies on the shape and path of small air bubbles rising in clean water." *Physics of Fluids* 14:49–52.

Ybert, C. 1998. "Stabilisation des mousses aqueuses par des protéines." Ph.D. thesis, Université Louis Pasteur, Strasbourg, France.

Ybert, C., and Di Meglio, J.-M. 1998. "Ascending air bubbles in protein solutions." *European Physical Journal,* Series B, 4:313–319.

ACKNOWLEDGMENTS

The author expresses gratitude to everyone on the wonderful team at Princeton University Press for their kindness, skill, and rigor, and especially to Ingrid Gnerlich, the physical sciences editor, for being so enthusiastic at the time this bubble book was just the spark of an idea and for her constant and precious help throughout the entire writing process.

INDEX

Abbey of Hautvillers, 9, 11
adsorption, 66
aging, 25–26, 27, 62
alcohol, 20, 21, 22, 24, 65; types of, 102
aldehydes, 102
anechoic chamber, 105, 105n
Antoinette, Marie, 32
Archimedes' principle. *See* buoyancy
avalanche behavior, 105–106

beer, 7, 55, 65, 76, 101, 107; and "cleansing" effect of bubbles in, 76. *See also* bubbles, comparison between beer and champagne
blending (*assemblage*), 23–24
Bordeaux (France), 141
Botrytis cinerea, 141, 141n
bubble cap, 87, 108, 120, 123, 125; critical thickness of, 87; deformed, 120n; disintegration of, 87, 89, 92
bubble cavity, 92
bubble collapse, 94, 97, 102, 105–106, 107–108, 120, 121
bubble nucleation, 41–42, 55, 79; and the critical radius of curvature, 41, 42, 42n, 55, 57, 78; nucleation sites, 42, 44, 45, 48, 51, 61, 121, 140
"bubble nurseries," 40–42, 44–45, 48
bubble production, 37–40, 48, 51; kinetics of, 54–55, 57
bubble trains, 33, 51, 54–55, 59, 74; stability of, 77–78, 80
bubbles, and champagne, 4–5, 7–8, 33, 97, 101, 103–104, 140–141; bubble acceleration, 63–65, 81, 83, 85, 85n; bubble deflation, 125, 127; and bubble rafts, 108; bubble shape, 85–87; and "bubble stars," 127; and dangers during fermentation, 14–15; flower shapes of, 107–108, 120–121;

bubbles, and champagne (*cont.*)
and fragrance, 102–103; frequency of, 51; and gravity, 64–65; increase of, 13–14; reduction of, 11–12; self-cleaning, 69, 71; size and growth rate of, 8–9, 59, 61–62, 64; and stability, 76–80; velocity of, 67, 69, 83. *See also* bubble cap; bubble cavity; bubble collapse; bubble nucleation; "bubble nurseries"; bubble production; bubble trains; bubbles, comparison between beer and champagne; buoyancy; carbon dioxide; surfactants
bubbles, comparison between beer and champagne, 71, 73–74, 76, 78–79
buoyancy, 48, 63, 66, 86
Burgundy (France), 10, 11
Bush, George W., 134

carbon dioxide, 8, 11, 14, 20, 22, 24–25, 34, 44, 64, 65, 125, 133–134; and bubble growth, 59, 61–62; and bubble production, 37–40, 51, 54, 57
cellar master, 23
Champagne (France), 9–10, 11, 12, 13, 137–138, 139–140, 142
champagne bouquet, 26, 27
champagne, 4–6, 22; aging of, 25–26, 27; fizzing sound of, 104–107; flavor of, 26, 121; history of, 9–17; levels of sweetness of, 29. *See also* bubbles, and champagne; *Méthode Champenoise*
chardonnay, 20
Charles II, 12
collerette, 8, 125
corks, 14–15, 29, 33–34, 62
coupe. See goblet
Culick, Fred E. C., 87, 89

dégorgement (disgorging), 28, 35
Dom Pierre Pérignon, 11–12, 13–14, 17; and the art of blending, 16
dosage, 29, 35
drafting, 81, 83
drafting, kissing, and tumbling scenario, 81
drag, 63, 67
drop impact, 96–97
droplets. *See* jet drops
Duteurtre, Bruno, 3

ebullition, 54, 54n
Edgerton, Harold, 4, 96, 97
effervescence, 2, 7, 8, 13, 62, 121
England, as a grapevine growing region, 141–142
enhanced greenhouse effect, 133–134, 136
enologist, 23–24, 24n
enology, 24n

Epernay, 3
ethanol. *See* alcohol

fatty molecules, 123, 125
fermentation, 10–11, 20–21; dangers of exploding wine bottles during, 14–15; first fermentation, 20–22; second fermentation (*prise de mousse*), 24–25; and yeast, 21–22. See also *Méthode Champenoise*
fizz, 11, 103–104; sound of, 104–107
flavors, of wine, 26, 62, 121
flute, 8, 31, 33, 54, 86–87, 108, 123, 140
French Method, 19

Gay-Lussac, Joseph Louis, 21
glass technology: in England, 15–16; in France, 14–15
global warming, 133, 137–139, 142; and climatic weather patterns, 136–137; effect of on champagne making, 137–138
glucose, 22
glycoproteins, 123
goblet, 31–33
grape must, 20
grapes, 23, 137–138, 141; acidity of, 23, 138–139; noble grapes, 20, 20n; types of, 20, 20n
greenhouse effect. *See* enhanced greenhouse effect
gushing (*gerbage*), 34

Henry's law, 25, 25n, 37
homogeneous nucleation, 42

James I, 15
jet drops, 94, 101, 103–104, 107, 121

Keeling, Charles Dave, 134
kissing, 83
Kyoto Protocol, 134, 136

Lemaresquier, Hervé, 97
liquid jet, 104
Louis XIV, 13, 16
Louis XV, 16–17, 32
Louis XVI, 32
lumen, 44, 51

Mansell, Robert, 15–16
Merret, Christopher, 12–13
Méthode Champenoise, 19; aging in, 25–26, 27, 62; blending (*assemblage*), 23–24; *dégorgement* (disgorging), 28, 35; dosage, 29, 35; first fermentation, 20–22; riddling (*remuage*), 26–28; second fermentation (*prise de mousse*), 24–25

Méthode Traditionelle, 19
Moët, Claude, 17
Moët & Chandon, 3–4, 17, 44; Dom Pérignon blend, 17

nocieptors, 101–102
nonclassic heterogeneous nucleation, 42

organic acids, 102
Ostwald ripening, 127
oxygen, 22

Paris, 9, 10
Pasteur, Louis, 22
pinot meunier, 20
pinot noir, 20
Pompadour, Madame de, 31–32
proteins, 123, 140–141

Rayleigh-Plateau instability, 94
Reims, 9, 17
riddling (*remuage*), 26–28
Royal Society of London, 12
Ruinart, Nicolas, 17

Shafer, Neil, 73
shear stresses, 120
sommelier, 24n
sparkling wines, 7, 12, 14, 19, 24, 62, 65, 101, 107
Stevenson, Tom, 12n
sugar, 13, 20, 24, 29
surfactants, 65–67, 69, 74, 76–77, 121, 123, 125, 140

Taylor, Geoffrey Ingram, 87, 89

Van der Waals forces, 40, 40n
Vandewalle, Nicolas, 105
volatilization, 54n

wake instabilities, 79, 79n, 81
Walt Disney Studios, 96
white noise, 106
wine waiter. *See* sommelier
wines, 11; red, 31; white, 31. *See also* flavors, of wine; grapes

yeast (*Saccharomyces cerevisiae*), 20, 21–22, 22, 24, 25. *See also* riddling
yeast autolysis, 26

Zare, Richard, 73

SAIT LIBRARY
10205338